Research on the Communication Effects and Mass Media Credibility in China

Guoming Yu

Research on the Communication Effects and Mass Media Credibility in China

The Foundational Theory, Evaluation Methods and Empirical Analysis

中国财经出版传媒集团
经济科学出版社
Economic Science Press

Springer

Guoming Yu
School of Journalism and Communication
Beijing Normal University
Beijing, China

Translated by
Yaozhong Lyu
Qingdao University of Science
and Technology
Qingdao, China

Ceng Zeng
Qingdao University of Science
and Technology
Qingdao, China

Donghua Li
Trinity College Dublin
Dublin, Ireland

ISBN 978-981-19-6241-7 ISBN 978-981-19-6242-4 (eBook)
https://doi.org/10.1007/978-981-19-6242-4

Jointly published with Economic Science Press
The print edition is not for sale in Chinese Mainland. Customers from Chinese Mainland please order the print book from: Economic Science Press.

Translation from the Chinese language edition: "中国大众媒介的传播效果与公信力研究——基础理论、评测方法与实证分析" by Guoming Yu, © Economic Science Press 2009. Published by Economic Science Press. All Rights Reserved.

This Springer imprint is published by the registered company Springer Nature Singapore Pte Ltd.
The registered company address is: 152 Beach Road, #21-01/04 Gateway East, Singapore 189721, Singapore

Main Members of the Research Group

Zhang Hongzhong
Jin Yi
Ding Hanqing
Ou Ya
Zhang Yan
Han Xiaoning

Preface

The book represents a systematic theoretical exploration of media credibility. By defining the concept of media credibility and related ones both in Chinese and English, it initially discusses on dimensions of judgment and influencing factors of media credibility from the cross-region differences and historical changes of regimes and social transition. It proposes new ideas about analytical theories stemming from the Western society, and brings forward an original theoretical hypothesis of judgment dimensions of media credibility based on the reality of China as well as a generative pattern of media credibility. A comprehensive analysis probes into influences of public credibility on media functions. By investigating five aspects of media credibility, i.e. definition of concepts, dimensions of judgment, analysis of influencing factors, generative mechanism and influences on communication effects, the study sheds light on two issues related to China's current situation. First, it demonstrates the influence of declined media credibility on government image affected by social systematic factors. Second, it shows potential changes in media credibility in the process of marketization.

Based on research, encapsulation of measurement theories and approaches related to media credibility and numerous empirical case studies of international academia in the past one hundred years, the book develops an index to quantify mass media credibility of China. It is important to probe into the actual condition of Chinese media credibility and discuss practical approaches to enhancing such public credibility. To accomplish this goal, the study establishes a measurement system of media credibility suitable for Chinese social conditions. A precise understanding of actual condition of media credibility provides a set of practical solutions and suggestions for the development of media credibility.

Thus, this book focuses on distinct characteristics of China's media credibility and the development of relevant evaluation indexes. It highlights two fundamental issues of measurement approaches: (i) probing into basic factors for media credibility, and (ii) formulating an assessment scale of media credibility. Based on survey data, the study analyzes the importance of various assessment benchmarks for measuring media credibility and characteristics of credibility assessment.

Finally, an assessment scale of media credibility is created by screening and analyzing measurement indexes with statistical methods including exploratory and confirmatory factor analyses, and reliability and validity testing. The scale is of practical value.

Beijing, China Guoming Yu

Contents

Part I
A Theoretical Study on Mass Media Credibility

Chapter 1
Definitions of Key Concepts

1.1 The Meaning of Credibility

1.1.1 Interpretation of Credibility

"Credibility (公信力 in Chinese)" is a new word in Chinese, and it is not found in Chinese reference books such as *Ci Hai*, *Modern Chinese Dictionary* and *Xinhua Dictionary*. Zheng Yefu, a Chinese sociologist, believes that the meaning of the Chinese word "信" in dictionaries can be classified into the following sequences: (1) letters and messages; (2) the messenger (a person who delivers letters); (3) credentials and deeds; (4) honesty; (5) credit; (6) believe; (7) trust; (8) indeed; (9) let it go. Among them, "honesty", "credit", "believe" and "trust" are abstract words. Of these four words, "believe" and "trust" are words that express subjectivity and judgment of the subject. "Honesty" and "credit" show attributes of the observed object and are words expressing objectivity, such as judging whether the other party has credit or is honest. "公" has two meanings: one is the public of the subject; the second is the public of the object (Zheng, 2001: 8–9).

The meaning expressed in the concept of credibility can be summarized as shown in Table 1.1.

The following analyses the vocabulary "信" that express the subjectivity and objectivity.

Basic meanings of the word expressing "believe", "confidence", and "trust" which shows subjectivity are the same, and there are only some differences in a linguistic sense. The word meaning "trust" is a special concept in the field of sociology and economics. This book uses the word "trust" when discussing subject attributes in the subject-object relationship of credibility.

Similarly, basic meanings of the word expressing "honesty" and "credit" which shows objectivity are also the same, and carry some linguistic differences. "Honesty (诚实)" and "credit (信用)" expressed similar meanings in ancient China. Cheng Yi thought that "honesty follows trust and trust implies honesty", and Zhang Zai said, "honest at heart is called trust" which means that when dealing with relations we

G. Yu, *Research on the Communication Effects and Mass Media Credibility in China*,
https://doi.org/10.1007/978-981-19-6242-4_1

Table 1.1 Basic dimensions of measuring the concept of credibility

"公"	
Subject: The public	Object: Public authority or public institution
↓ "信"	↓
Words that express subjectivity	Words that express objectivity
Xiangxin (believe) Xinren (trust) Xinlai (confidence)	Chengshi (honesty) Xinyong (credit)

should live up to our words and strive to fulfill our promises and obligations. In the disciplines of economics, law, etc., the term "credit" is used more frequently, and it is a concept category. Therefore, the book uses the word "credit" when discussing object attributes in the credibility subject-object relationship.

"力" reflects two meanings of the concept credibility. The first meaning indicates the relational attribute of the concept. The second meaning indicates its concrete meaning, that is to say, credibility is the ability of an object to win the trust of the subject.

1.1.2 Relationship Model of Credibility

According to the previous interpretation of the meanings of credibility, the relationship model of this concept is proposed here.

Credibility is a concept that implies the relationship. What is the relationship? The so-called relationship refers to a set of expectations two people have for each other based on their interaction (Littlejohn, 1999: 451). Credibility is a concept of the interaction between the subject and the object, i.e., the subject's (public) expectation of the object's credit, and the object's expectation is to win public trust. Take the credibility of mass media as an example. The public decide whether they trust the media or the degree of trust through knowing about the media credibility. Since the media needs to win public trust, they must acquire corresponding credit qualities. The relationship model is shown in Fig. 1.1.

In the above model, the second concrete meaning of this concept is analyzed.

For the credibility of mass media, the focus is on the object "mass media". From the perspective of the object, "力" refers to the first meaning "capacity". In brief, the credibility of mass media means the capacity of mass media to win public trust. The capacity of mass media depends on the trust of the public, and there is an interactive trust relationship between the media and the public. Therefore, trust is the starting

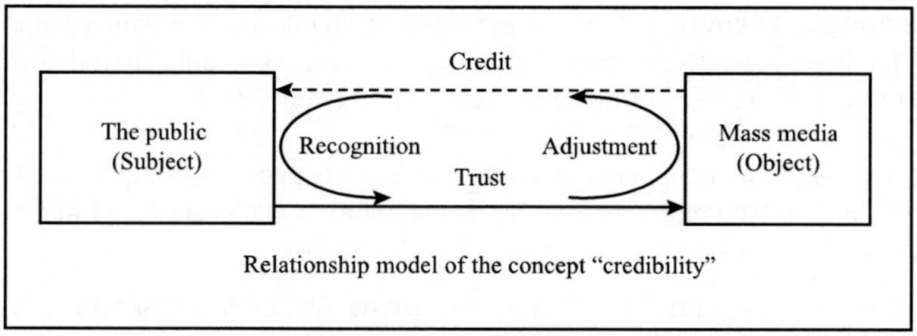

Fig. 1.1 The relationship model of the concept "credibility"

point of the credibility, showing that the object's credit is an evaluation element of credibility.

1.1.3 The Logical Starting Point of the Trust

The model of credibility shows that "trust" is the logical starting point for the concept of credibility. What is the logical starting point? According to the philosophy of logic, the starting point of logic, also called the beginning of logic, is the starting category of a theoretical system. It is the beginning of a hologram, which is derived from an entire theoretical system. In other words, the logical starting point is like the embryo of a theoretical system.

In this book, each issue on the credibility theory of mass media begins from the logical starting point of "trust." Let's look at the conceptual meaning of "trust".

1.1.3.1 The Meaning of "Trust"

The English word "trust" has various interpretations in different social disciplines such as sociology, economics, political science, and anthropology. There are two overarching definitions.

(1) Many scholars define "trust" from the perspective of subjective behavior and believe that "trust" is a psychological behavior.

John Donn considers "trust" as both a human passion and a modality of human action. To some extent it is a strategy for handling the freedom of other human agents or agencies (Donn, 2003: 89).

Coleman defined "trust" as "making a risky decision on whether or not to take the other's actions" by making predictions about others' behavior in the future (Kramer, 2003: 12).

According to Fukuyama, "trust" is the expectation that arises within a community of regular, honest, and cooperative behavior, based on commonly shared norms, on the part of other members of that community (Fukuyama, 2001).

Diego Gambetta defines it specifically. "Trust" is a particular level of the subjective probability with which an agent assesses that another agent or group of agents will perform a particular action, both before he can monitor such action and in a context in which it affects his own action (Gambetta, 2003: 270–271).

(2) Some scholars define "trust" from the perspective of function. Zheng Yefu, a Chinese scholar, thinks that "trust" is an attitude and believes that someone's behavior or the order around them meets their wishes. It is manifested in three kinds of expectations, the order of nature and society, the obligations to partners and the technical capabilities of certain roles. "Trust" is the medium of exchange (Zheng, 2001: 19).

Based on the definitions from the above two perspectives, this book believes that "trust" is the subject's expectations of the object's future behavior. This expectation affects the subject's next action on the object.

There are four features of "trust" in this sense.

(1) Time difference. 信任 (the Chinese word means trust) comprising of "信" and "任" is the judgment and behavior in a chronological order. "Yang Huo" of *The Analects* says, "If you are generous, you will win all. If you are trusted, you will be employed." "The Robber Kih" of *Chuang-Tzu* says, "Without righteous course, you will not be trusted. Unless you are trusted, you will not be employed. If not employed, you will not acquire gain." Chuang Tzu stated the premise and result of trust: virtue is the premise of trust which is the premise of employment and the final result is benefit. The premise of trust is "virtue", and its final result is "benefit".

Zheng (2001: 19) summarized that the first nature of trust relationship is time difference and asymmetry. He believes that action and fulfillment must follow promises and agreements. There is a time lag between words and actions, commitment and fulfillment. There must be some asymmetry between the trustee and the trusted.

(2) Uncertainty. This is a prerequisite for trust, and it emerges under a state of uncertainty. With certainty, there is no specific way to handle risks, and there is no such a thing as trust (Zheng, 2001: 19). Diego (2003: 270–271) thinks that trust is particularly relevant in conditions of ignorance or uncertainty with respect to unknown or unknowable actions of the other. In this respect, trust does not concern future actions in general, but all future actions which condition our present decisions.
(3) Subjectivity. Trust is a psychological activity of the subject, and it is an evaluation and judgment of the subject to the object. Just as the last point of the three natures of the trust relationship summarized by Zheng (2001: 19), trust is a subjective tendency and intention for there is no sufficient objective basis. The

subject's follow-up behavior to the object and the way it takes depends on this subjectivity of trust, as Diego (2003: 270–271) postulates that our own actions are dependent on that probability.

(4) Valve point. Valve point means that trust is like a valve that controls an action. Only when a certain valve point is reached can an action take place. Therefore, different degrees of trust correspond to different actions. Trust is better seen as a valve point, located on a probabilistic distribution of more general expectations, which can take a number of values suspended between complete distrust and complete trust, and which is centered around a mid-point of uncertainty (Gambetta, 2003: 270–271).

1.1.3.2 Extension of Trust

At present, the research on the extension of trust is relatively broad. The subject and object of research cover individual to individual, individual to institution, and group to institution, etc. This book addresses this issue from the perspectives of the extension of the subject and the object.

(1) For different subjects of trust, *the extension of trust includes trust on the* individual-level and group-level.

Trust at the individual-level is basic. Jacint Fordana, professor of Political and Social Sciences Department, Universitat Pompeu Fabraa in Barcelona, Spain, believes that social and political theories should be built on the basis of individual intentions, so that micro factors can be found in every theoretical explanation. In the field of social sciences, the fundamental element is the individual behavior (Cao, 2003: 200). As a research category of social sciences such as sociology and economics, individual is the basic research unit of trust.

Trust at the individual level consists of two aspects, one of which is the interpersonal trust, that is, individual to individual. The interpersonal trust is an important part of research on trust. Since 1970s when trust became a specialized research topic in sociology, the interpersonal trust has been discussed in works such as Fukuyama's *The Social Virtues and the Creation of Prosperity*. The other aspect is the individual's trust in things (such as institutions, etc.), that is, individuals to things (such as institutions, etc.). These things are managed by people. It is the relationship between individuals and individuals, and individuals and groups, but they are not directly connected. They are connected through external things. According to David Good's understanding of trust in the *Oxford English Dictionary*, the first substantive definition of "trust" is "confidence in or reliance on some quality or attribute of a person or thing, or the truth of a statement". A later variant stresses the economic usage, defining it as "confidence in the ability and intention of a buyer to pay later for goods supplied without present payment". Under either these definitions, or the others offered in this volume, trust is based on an individual's theory as to how another person will perform on some future occasion, as a function of that target person's current and previous claims, either implicit or explicit, as to how they will behave (Good, 2003: 365).

Trust at the group level is another component of subjective extension, such as public-to-object, receiver-to-object, etc. Trust, as a theory, is based on individual intentions, and its basic unit is an individual, but it does not mean that its analysis must be limited to individuals. Since individuals have many different sets, such as groups and the public, the analysis can be done at different levels. Some scholars think that trust exists widely at different levels of analysis. It exists at individual level, within groups or organizations and between organizations, and even at institution level, such as public trust in the political system. For a set of individuals (such as a group or a whole organization), the working definition of trust is the average level of trust among group members (Mishra, 2003: 365).

(2) *For extension of the object of trust, it is typical to divide trust into personality trust and system trust according to* the theory of trust structure.

Like the subject's trust extension, it can be divided into different kinds of trust extensions according to different objects. Anthony Giddens, British sociologist, summarizes the trust structure between two kinds of objects from the perspective of sociological trust structure theory. One is personality trust, that is, personal trust, which also refers to the interpersonal trust discussed above; the other is impersonal system trust, such as subject to government, subject to media, and subject to currency.

Scholars think that personality trust is a basic feature of the traditional social structure. It originates from human nature and instinct. Giddens (1991: 4) argues, trust is a crucial generic phenomenon of personality development and it is directly linked to achieving an early sense of ontological security. It is based on the relationship between specific interpersonal kinship and familiarity, and is the most basic trust. System trust is a main feature of the modern society. The traditional social structure is based on the community of acquaintances, so the trust is interpersonal trust among acquaintances. The biggest feature of modern society is that it goes beyond acquaintances and the trust is built on an abstract system...Although modern social life cannot exclude interpersonal trust, it is increasingly dependent on system trust (Zheng, 2001).

Therefore, personality trust and system trust embody a relationship between tradition and modernity. Zheng (2001) thinks that traditional personality trust and modern system trust are in an inheriting, corresponding and coexisting relationship. If we are required to compare the importance of inheritance and correspondence, it is undoubtedly the inheritance. A nation that has wiped out its traditional spiritual resources will not have a bright future. If the traditional organizational resources no longer exist, the expected result is the extremely long rest, growth and recovery in culture and civilization. Today sociologists believe that personality trust and system trust influence each other. At present, lack of trust and trust crisis have attracted people's attention and uneasiness of our social life, so it is far more than enough to just see the lack of trust at the individual level. And many personal trusts cannot be separated from the system trust.

1.1.3.3 The Logical Chain of the Concept of Credibility

From the above analysis of trust extension, it can be seen that from the perspective of the object, system trust is a logical link of the concept of credibility, that is, the logical chain of the concept is trust to system trust and to credibility.

Some scholars think that "system trust is actually a kind of credibility" (Sun, 2002). In fact, there is a lexical difference between system trust and credibility. First, system trust does not define the scope of the subject of trust. It can be the trust of individuals, groups or the public on the system, while credibility clearly refers to the public. Second, there is no definition of "力" in system trust, that is, characteristics of the concept and "能力 (meaning ability)" to express object attributes. But system trust, like credibility, expresses attributes of objects, and its objects are the government, the media, and so on. System trust is a logical loop of the concept of credibility. Similarly, the credibility of mass media is as much a feature of modern society as system trust.

1.1.4 Concept Evaluation—Credit

1.1.4.1 The Meaning of Credit

There are four explanations of "信用 (meaning credit)" in *The Modern Chinese Dictionary*: (1) the status of trust that can be obtained by fulfilling promises agreed with others; (2) it is not necessary to provide material guarantee and can be repaid on time; (3) it refers to bank loans or commercial credit sales and purchases; (4) (formal) trust and appointment. The interpretation item (2) and (3) are used commerce.

Ci Hai lists three interpretations of credit: (1) the use of trust; (2) to keep the promise and practice the contract, so as to gain the trust of others; (3) special forms of value movement.

The credit, often used in economics, is an important concept in economics. The following is definitions of credit given by economists to review its academic meaning.

Credit has its broad sense and narrow sense. In the broad sense, credit refers to the ability to fulfill the contract based on honesty and trustworthiness established among parties involved in economic activities, that is, what we usually say "having credit", "honoring credibility" and "a promise is gold". The narrow sense of credit under modern market economy refers to the ability to fulfill the payment or repayment commitments made by the trusted party at a specific time (including the ability to perform various types of economic contracts).

According to Wu Jinglian, a famous Chinese economist, credit refers to the ability to build trust based on the trustee's payment or repayment of promises within a specified period of time. It is the ability to make the latter acquire goods, services or money without paying cash. Credit is an essential element of making market transactions develop (Ma et al., 2001).

Credit is an attribute that expresses the object (observed object). This attribute is established in the interaction with the subject (the observer), and it is expressed as the subject's expectation of the object's future actions. This expectation has become an ability of the object itself, and this ability can be quantified as the degree of credit, which indicates the status of credit worthiness of credit stakeholders (Song, 2003).

1.1.4.2 The Relationship Between Credit and Credibility

Credit is a word to refer to the object, and it is an important evaluation content of credibility. Scholars in some disciplines such as politics and ethics have discussed the relationship between credit and credibility, and they think that credibility is the degree of public trust in object credit. Here the difference between the two is discussed in a paper in Administrative Science:

Concepts of government credibility and government credit are closely related but different. The government credit refers to the government's fulfillment of its commitment to the public, and it is an important indicator of a responsible government under conditions of modern democracy and rule of law. The government credibility reflects the government's credit ability, and it displays the extent to which citizens have trust in government actions. The strength of the government credibility depends on the abundance of credit resources owned by the government. Such credit resources include both ideological and material, as well as the specific image of the government and its staff in the eye of citizens and so on (He et al., 2002: 44).

From this perspective, credit is the content of the object that the public evaluates in the concept of credibility. Credibility in different disciplines have different target credit evaluation. As mentioned above, the government credibility is the public's evaluation of the government credit. The credibility of news media is the public's evaluation of news media's credit. The credibility of non-profit organizations is the public's evaluation of their credit. Economic organization credibility is the public's evaluation of their credit.

1.2 The Meaning of Mass Media Credibility in English

There has been no consensus on the English counterpart of "媒介公信力 (the media credibility)". Some think that the English vocabulary for "公信力 (credibility)" is public trust, and some agree that it is credibility. American scholars of Communication focus on credibility. Some translated this word as public trust and some translated it as reliability in Chinese. The meanings of these words are different, and because of these differences, it is easy to cause confusion in research.

Therefore, this section attempts to sort out this issue and focuses on the connotation of credibility.

1.2.1 Two Viewpoints of Credibility

1.2.1.1 The Attribute and Relational Variable of Credibility in Research

In the late 1950s and early 1960s, the study of media credibility became an important research area in the United States. After decades of continuous research, much progress has been made on the measurement and the judgment dimensions of credibility. There are two kinds of definition of credibility.

(1) The "attribute" variable. Scholars who hold this view think that credibility is perceived as an attribute of the media (or message source), which depends on the media itself and has nothing to do with the receiver.

As Gunther (1992: 148) pointed out in his analysis of credibility, the most common explanation of credibility refers to media characters. Ben Bradlee, an editor from *Washington Post* thinks that "The credibility of a newspaper is its most precious asset, and it depends almost entirely on the integrity of its reporters" (1981). Indeed, in research on mass media, credibility has been defined primarily as an attribute of the source.

Research between the 1950s and 1960s centered on the media, such as investigating the credibility of media and comparing the credibility of different media.

(2) The "relational" variable. Credibility is the relationship between the receiver and the media. Credibility is shaped by the receiver's perception. The receiver is important for credibility, so it is not solely depended on the media itself.

In the mid to late 1960s, scholars began to rethink the meaning of credibility. Jacobson (1969: 20) in the research on message source of credibility argued that the theoretical orientation in the credibility studies of the Yale group was a starting point. More recently, Berlo et al. (1969) made considerable advances, conceptualizing credibility as a relational rather than a property term, conceiving of it as part of a broader judgment of the source and manipulating it in terms of receiver perceptions. In the past credibility has typically been assumed to be unidimensional, dichotomous (either high or low) and specifiable in terms of objective characteristics of the source, such as social role. Such a stipulation has implied that the variable is a static attribute of a source, rather than a receiver evaluation which is subject to change; such an approach has precluded consideration of the variable as a dependent outcome of communication.

From the perspective of the receiver credibility is a word that expresses the receiver perception. For example, Gunther (1992: 152) thinks that media credibility is a receiver assessment, not a source characteristic, and therefore variables predicting credibility are more likely to be associated with the receiver than with the source. In fact, all scholars who study credibility describe it as being perceived and multidimensional. Credibility is a perceived quality; it doesn't reside in an object, a person, or a piece of information (Fogg, 2002).

From attribute to relational, scholars no longer consider credibility as an objective attribute that depends solely on the media (or the source of the message). It is examined in the process of dissemination and realized through the receiver's perception, which is affected by different status of the receiver. For example, the same message source is published on the same medium and credibility is the same to different receivers according to the early views, for it is an objective characteristic of the media and is relatively constant. However, if credibility is considered as a relational and a product of communication, it is the result of the receiver's gender, age, political ideology, media use, and media dependence. Thus, different receivers have different media credibility. Here, the latter lies on the receiver.

This book examines the concept of credibility from the perspective of relational. Credibility corresponds to the concept of credibility in Chinese. It is a relational concept. The attribute of the media is determined by the receiver's trust in the process of communication, rather than exclusively depends on the media itself. The relational variable studies the media credibility from the perspective of the receiver, broadening the meaning of credibility. Scholars have agreed that credibility is a perception of the receiver, which is an important feature of credibility.

1.2.1.2 The Relationship Between Credibility and Public Trust

According to the English explanation of credibility, it is a concept representing quality and ability, and its concept is the quality, or the quality capability.

Zhou (2004b), a scholar at the University of Iowa in the United States, thinks that there is a distinction between credibility and public trust. He argues: "Credibility is defined from the perspective of the communicator. We say that a certain message source, some news content, or a certain media is more credible because these sources, content, or some characteristics inherent in the media make the receivers feel that they have attributes of credibility. The concept of public trust is defined from the perspective of the receiver. The receiver voted 'trust' for certain media, message sources, and content based on a variety of factors. Some factors may include the receiver's cultural environment, personal experience, education level, and their media use. The credibility of the message source, content, and form plays an important role in choice of trust. Media workers recognize that the receivers themselves cannot control the factors, but the characteristics of the message are determined by the communication workers. Therefore, credibility has become the focus of research."

Zhou Shuhua made a clear distinction between the definitions of credibility and public trust in Chinese language. He thinks that the concept of credibility in Chinese is believability while the public trust corresponds to "public credit".

Zhou Shuhua defines credibility based on the connotation of "the attribute" from the perspective of the communicator, but not from the development connotation of credibility. From the connotation of "relational variable" the study of credibility is actually the definition of public trust, as defined from the perspective of the receiver, influenced by the receivers' education level and environment.

Public trust is literally translated as "public credit". Trust is the logical starting point of the concept of credibility and includes the meaning of public trust. Public trust does reflect two meanings of "力 (capacity)" of credibility in English.

Therefore, credibility is used in this book.

1.2.2 Trust, Credit, Faith, Confidence and Believe

Let's look at several other English words meaning "信".

Zheng Yefu made a semantic distinction between two English words of Chinese word "信任". He thinks that a word similar to "trust" is "confidence". These two words are synonyms for "Xinren" in English. Other scholars think that the main difference between the two is in the degree, and the former is strong in the degree of trust. Some think that "confidence" refers to trust in a system, and "trust" refers to the attitude towards boundaries and gaps of a system. Luhmann argues that "confidence" is connected with danger in things as non-selective, while "trust" is bound to risk and subjective choice (Zheng, 2001: 12).

It is difficult for English scholars to agree on the distinction between "trust" and "confidence". The basic meanings of the two words are the same, but there is a difference in the linguistic sense. "Confidence" is stronger than "trust". The set of synonyms to which "trust" belongs is usually confusing. Faith, trust, and confidence all express belief (Hart, 2003: 232). In addition, the English word for "相信" is "believe". These words are all of subjective attitudes.

For "信用", the corresponding English word in economics is "credit". For example, Law Merchant system is equivalent to the credit reporting bureau in modern society (Zhang, 2001c). The credit refers to the ability of a receiver to obtain goods or services before paying, based on the trust of the creditor in the promise of the receiver to pay (Wu, 2005).

1.3 The Definition of Media Credibility

According to the above discussion, media credibility is defined from aspects of the logical starting point, connotation, extension and communication characteristics of the concept.

(1) The logical starting point of media credibility is "trust". What is trust? This book agrees that "trust" is the subject's expectation of the object's future behavior, and this expectation affects the subject's next action on the object. Starting from trust, the following chapters analyze the judgment dimension of media credibility, influencing factors, generative mechanism and the impact on the function of media.

(2) What is media credibility? This book holds a view that media credibility is the ability of the media to win public trust in the interaction between the public and the media. From the perspective of the media, its content is the credit of the media, and credit is the evaluation of the credibility of the media. This evaluation is based on the public's perception. Credibility is conceived as part of a broader judgment of the source and operationalizing it in terms of receiver perceptions (Berlo, 1969: 563–576), that is, the media credibility refers to a kind of subjective behavior of the public.
(3) The extension of the media credibility consists of the credibility of the media as a whole, the credibility of media channels, the credibility of message sources, the credibility of the press as well as the credibility of journalists. In the United States, scholars have studied the credibility of media channels and the credibility of message sources. Some scholars think that credibility studies have focused on message sources and transmission channels (Kiousis, 2001). Chinese scholars have also discussed the credibility of the overall media and the credibility of the press. The media credibility refers to the degree of the trust the audience place in the news media. The media must rely on fair and objective reports and real news to gain the trust of the audience. Otherwise, it is difficult to win the trust of the audience (Chen, 2003). The media credibility refers to the social media credibility, that is, the degree of the public positive evaluation and praise of the media. It is the media institution that obtains trust from society and the public, maintains its good social image and strive for its own survival and development (Zheng, 2003a). The media credibility refers to creditworthiness, authority and influence of society and audience in the long-term development process of media. The media credibility is the social power, or social influence, and media capacity, which is generated by public trust. The extension of the media credibility includes media channels, message sources, organizations, and practitioners, as well as the overall credibility of the media.
(4) The credibility of the media is established during the communication process and is the product of the communication process. So, where is the credibility of the media in the communication process? Some scholars say, trust can be seen as a result rather than a precondition of cooperation (Gambetta, 2003: 265). Other scholars argue that trust is the beginning of a cooperative relationship. This book believes that the media credibility is the result of previous communication campaign and affects next one.

1.3.1 The Contractual Relationship Between the Public and Mass Media

1.3.1.1 "Non-social Information Needs" and "Social Information Needs"

The information needs of the people come from their environment. The nature of information needs is quite different. Some information needs are only for entertainment, such as gossip news, interesting stories, TV series and other media information. Although the entertainment function of the mass media has become more and more prominent in recent years, mass media is not a purely commercial organization. It is a kind of "social public instrument". It is widely acknowledged that mass media should promote the "public interest". Therefore, the people have placed expectations on the media to monitor the environment, supervise the government, inherit culture and maintain social order. These social expectations related to the public interest constitute another type of information needs for the media. We refer to the information needs related to the public interest as "social information needs" and the rest of needs that are not related to the public interest are called "non-social information needs." The public groups that put forward the "social information needs" are connected to public interest demands and have a certain public spirit. In this sense, the media audiences themselves have also become the public.

1.3.1.2 "Non-social Contract" and "Social Contract"

The contract is used to stipulate the responsibilities, rights, and benefits of the contracted parties, to ensure mutual benefits and to restrict each other. The relationship of interaction between media and public is actually a contractual relationship based on their respective responsibilities, rights, and benefits. The differences of people's media information needs determine different contents of the contract. One of the important differences lies in what methods the people use to ensure that their different information needs are satisfied.

The media's satisfaction of the people's "non-social information needs" is similar to a simple commodity exchange relationship, and the contract between the people and the media can be called a "non-social contract". The content of the contract can be briefly described as the audience pays money (or attention) for media products to meet their "non-social information needs". The media provides the audience with media products that meet the needs of the audience and thus get money (or attention). When the media cannot meet the audience's "non-social information" needs, the means of punishment and supervision adopted by the audience is no longer buying or contacting the media, which directly affects the economic benefits of the media. It is an economic punishment for the media.

For the information needs related to the public interest, "social contract" is signed between the media and the public. What is different from the "non-social contract" is that the means of punishment for not satisfying the public's "social information needs" punishment include the economic punishment of not buying or contacting the media and the "public trust punishment", which is to distrust the media.

Therefore, the essence of the media credibility is the rational reflection of the degree to which the public's social expectations (social information needs) for the media are met.

1.3.1.3 "Economic Punishment" and "Trust Punishment"

From the above discussion, it can be found that the public's constraints on the media's fulfill their social responsibilities, (i.e. fulfilling the public's social expectations), including "economic punishment" and "trust punishment". Economic punishment is a relatively quick and direct restraint. Economic punishment can be quickly shown through the decline in median circulation and ratings, thereby urging the media to better meet the public's social expectations. If it is in a freely competitive media market, those media that do not respond to the needs of the public will be driven out. Economic punishment is directly related to the economic interests and development of the media and is a more powerful means of restraint. However, "economic punishment" does not exclusively guarantee that the public's social expectations can be achieved. The media can also receive economic support if it meets the "non-social expectations" of people, thereby maintaining the survival and development of the media. Sometimes, "non-social information needs" of the media are contrary to the public interest, and the media is likely to take acts that harm the public interest for economic benefit.

Compared with the means of economic punishment, the public's trust punishment for the media has negative and lagging characteristics. Trust is a long process of psychological change, and this process has no obvious sign, and it cannot give an early warning to the media. And only when a considerable proportion of the people lose trust in the media, the effect of the trust punishment can be clearly demonstrated. Therefore, it takes a certain amount of time to accumulate the distrust ratio.

If the media system lacks economic punishment under the non-marketization media system, people's punishment for the media's failure to meet their social information needs will focus on the "trust punishment" for lack of early warning. Since "trust punishment" are lagging, it produces a much more hidden danger. The media may be unconscious before the crisis of trust appears, and the reduction or loss of public trust in the media will usually show up when the contradiction intensifies. At this time the media often pays a considerable price to restore trust. More important, it is not only a single media that suffers losses, but the entire media system and even the entire society may be harmed.

1.3.2 The Public's Social Expectations and Credibility of the Media

In the above discussion, "media credibility" is determined by two factors, "the public's social expectations of the media" and "the media's implementation of such social expectations". The social expectations of the public for the media and how these social expectations are better achieved through the media are the basis of credibility research.

1.3.2.1 Specific Content of Expectations for the Media

For the way mass media realizing social responsibility, the normative theories of Communication and Journalism provide several recommendations. McQuail (1994) made a series of summaries. He used concepts of freedom, equality and order as the core of the media performance, and provided a detailed analysis. 23 kinds of the performances are presented in Table 1.2.

There are many theories for what mass media should be. Theories, such as Libertarian theory, Marxist media theory and social responsibility theory have differences as well as similarities. The McQuail's summary in Table 1.2 presents opinions of several scholars.

For the study of "media credibility", the public is the decisive factor of "how the media should be". The public perception of the media (i.e. the public's social expectations for the media) is exactly the starting point for their judgment of the media credibility. Perhaps experts and scholars think some of the public opinions are superficial or even wrong. But they exist in real life. The public judge the media based on their performance. The public's social expectations of the media form the basis of the media credibility, and the public assessment of the media's social expectations reflects the credibility of the media.

1.3.2.2 Media Performance and Expectations

The public's expectation of the media is a psychological agreement to a large extent. These agreements are not only intangible, but the content of clauses and the evaluation criteria of each clause are related to the audience's personality, psychology, knowledge literacy, life experience and other factors. They are complex and changeable. The status of credibility is determined by the implementation of expectations. The degree of correspondence between the specific performance of mass media and the public's social expectations of the media are the determinant of media credibility.

There are three status points between the performance of the media and the psychological expectations of the public: failure to meet expectations, fulfillment of expectations and exceeding expectations. Different cases have different effects on the media credibility. When the public's expectations are realized, it has a positive

Table 1.2 McQuail's summary of media performance

Media performance	Freedom	Independence	Reliability	
			Originality	
			Critical stance	
	Equality	Access for senders	Open and/or equal	
			Proportional	
		Diversity for receivers	Diversity of supply	Choice
			Diversity of access and content	Change
				Reach
		Objectivity of the content	Neutrality	
			Fairness	
			Truthfulness	
	Order	Social	Solidarity	Empathetic
				Identified
			Pre-social influence	
			Control	Consensus
				Public order
		Cultural	Quality	Educational/scientific
				Artistic/Aesthetic
			Un-cultural	Teste
				In-decency
				Violence
			Identity	National
				Sub-cultural

impact on credibility. If the public's expectations are not satisfied, the credibility of the media is negatively affected. If the media can perform beyond the expectations of the public, it has a significant impact on the improvement of credibility.

To sum up, the examination of media credibility should consider the audience's expectations as the starting point and as the basis of the evaluation criteria of media credibility. Examining the credibility of the media from this perspective can address a similar problem. If people today were to read newspapers from the period of "Great Leap Forward", they would have thought that the news are not reliable. At that time, there was no specific survey data on people's trust in newspapers. However, we can infer that expectations were quite high. People have different expectations of the media in different periods. During the 1960s, the role of newspapers was similar to government documents in the eye of people. Therefore, "expectations" on newspapers are similar to expectations on "documents", whether the information is authentic is not the most important. The level of message source is the most important basis of trust. Exploring the media credibility from the perspective of the media expectations of the public also solves the problem that the media credibility evaluation standards cannot keep uniform in different historical, cultural and social backgrounds.

The logical relationships among the above-mentioned concepts of "contractual relationship among the public and the media", "media expectations" and "media performance" reveal the essence of the media credibility and its mechanism of occurrence. It can be shown in Fig. 1.2.

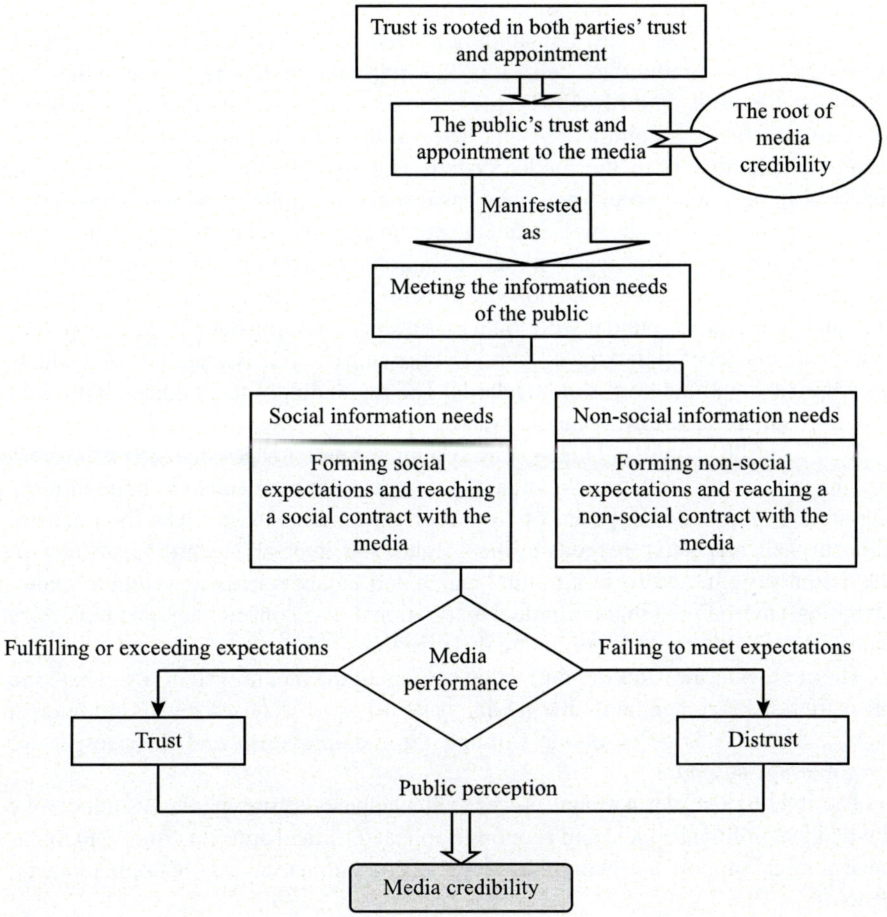

Fig. 1.2 Generative mechanism of the media credibility

1.4 Media Trust in Mass Media System

Regarding the significance and role of the media credibility, most researchers hold relatively consistent views that lack of media credibility will reduce circulation or viewing (listening) rates, that the government cannot convey official orders through the news media, and that the public cannot oversee the government through the news media (Lo et al., 1993). From the perspective of institutional economics and modern contract theory the following examines how the media credibility can achieve such effect, and what is the mechanism of its role.

A system is a set of rules, law-abiding procedures and ethical codes of conduct. It aims to restrict individual behaviors that pursue the welfare of the subject or maximize the utility (North, 1991). Mass media system (media mechanism/media system) is actually a contract system. The value of the media system is to make different stakeholders in the media system establish effective self-constraint and mechanism of mutual checks and balances through formulating various contracts.

National authorities, the media, and the public are the main participants and stakeholders in mass media system. Without institutional constraints, there must be excessive interest friction among stakeholders, affecting the efficiency of the media system. A media system with sound institutional constraints can bring benefits to each participant, and it is beneficial to the operation of the entire social system. Nothing can be accomplished without norms or standards. The regulation of the media system is an essential requirement for the media system.

During the development of the media system, the national power centers, the press and the public rely on their own strengths to meet their own and common interests. The media system is composed of formal or informal contracts. The three parties allocate their respective responsibilities, rights and interests according to rules of the media system, and form a mutual check and balance so as to obtain a media-operating-environment that is conducive to rational decision-making, and maximize their common interests and minimize their frictions of interest.

The content of the rules or contractual content of the media system (i.e. what kind of institutional arrangements the media system has) is determined by the interest that stakeholders hope to obtain through the media system and their respective "negotiating" strengths.

The interest that the national power center appeals to the media system can be divided into political appeal and economic appeal. Political appeal is simply to maximize political support, and economic appeal can be summarized as obtaining financial benefits.

The demands of the press can also be divided into two types: political and economic. Politically, it includes the realization of its own news, concepts, social status, etc. The economy relies on the media to obtain profits.

The public appeal for the media is to obtain information. The information needed is varying in different social backgrounds and historical periods. They can be divided into "non-social information needs" and "social information needs".

The "negotiating strength" of the three stakeholders is essentially the ability to restrict other stakeholders. It determines the ability of different stakeholders to bargain in the process of contract formulation (i.e. the media system arrangement), and the ability to bargain determines how much benefit subjects can finally get the right of benefit distribution from the media system arrangement.

In general, the national authorities have the full strength backed by the state (unless the state power is in crisis).

The strength of the media organization is the power of the media, and its power depends on the economics, political freedom and independence of the press. The vast majority of media in some countries (such as the United States) operate in a market-oriented way and have certain economic autonomy. The constitution protects its political independence, thus becoming the so-called "fourth power" with more economic and political power. The media institutions in some countries (such as Singapore) take market-oriented operation, so they have some economic strength but lack political power. Some presses are completely politically and economically subordinated to the national power centers. But no matter how powerful a press is, the media power is always under the control of the state. It cannot be compared with the power of the national authorities.

The people play a special role in the media institutions. They lack a strong and active means of safeguarding their own interests. From this perspective, the people have a weaker bargaining position in the process of media system arrangement. Whether it is the national power centers or the press, the realization of both political and economic demands depends on the public's acceptance of media message. Therefore, if the public refuses to accept media message, mass media system won't work.

Although the public lacks a strong and active means of checks and balances, they have a means of punishment that is significant to the overall media system—the exclusion of media message. This "exclusion" is reflected at the behavioral level without contacting the media, and at the psychological level, it is distrustful of the media. When the public believes that their interests (to obtain the required message) cannot be guaranteed by the existing media system, they exercise the "right to withdraw" as a punitive countermeasure by not touching the media or trusting the media. The use of the public "right to withdraw" result in the loss of the interests of all three parties, and the operation of the entire society will also be damaged.

At this point we may find a place for the media trust in the entire mass media system. The satisfaction of the public's need of information affects the public trust in the media, and public trust in the media as a function of mass media system affects the interests of other stakeholders (national power centers and the press). The media trust affects the operation of the entire mass media system and social system. Although the public trust in the media is a small part of the media system, the break in this link is very destructive. This is exactly why the media trust is important.

The logical relationship among the media trust, the three stakeholders, the media system and the social system can be represented in Fig. 1.3.

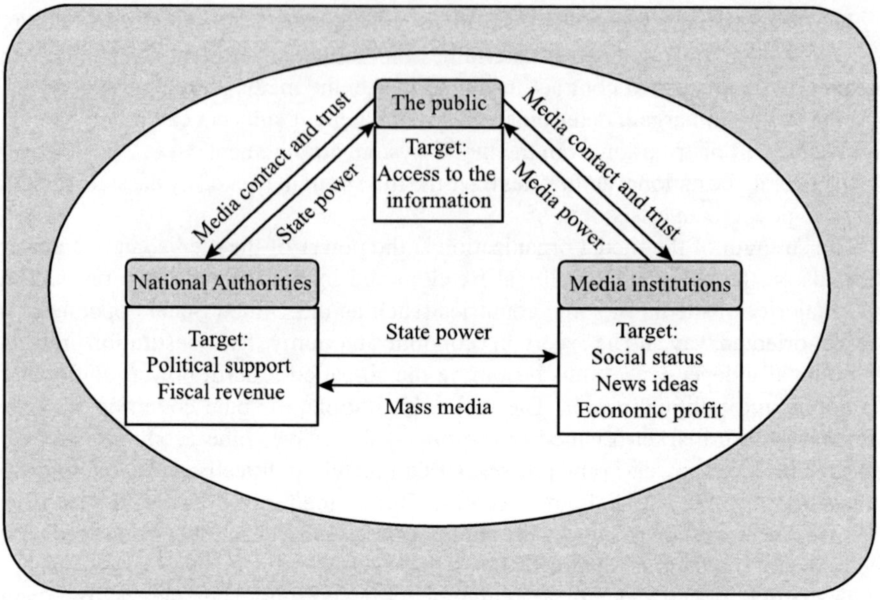

Fig. 1.3 Schematic relationship between the media credibility and the three stakeholders, the media system and the social system

1.5 Maintenance and Regulation of Media Credibility

In the media system where the media is fully market-oriented, the constraints on media social responsibility are mainly achieved by economic forces. Such constraints are based on a sound media market economic system. In a non-market-based media system, the public's economic constraints on the media are difficult to achieve because the media are essentially free of funds and economic pressure from bankruptcy. In this case, the public must use the "trust punishment" method. As discussed above, "trust punishment" has lagging and forbearing characteristics. It is not easy to detect before the collapse of trust, which may be a greater destructive power.

Different media systems have different strategies and means for maintaining the media credibility. In the environment where the media market is completely open, as in the United States, the social agreement between the media and the public mainly depends on economic means to maintain influence, and generally the government does not directly engage in the management of the media. The problem with this restriction is that the pursuit of economic interests by mass media is often irrelevant or even harmful to the public interest. In the United States, the threat of the media credibility mainly comes from the damage to the public interest by the media for economic benefits. Therefore, relying solely on the economy as a means of restriction cannot fully guarantee the media credibility.

The situation in China is different. The government is the representative of the interests of the people. The media is under the direct control of the government. The logical deduction is that the media represents the will of the people. The problem here is that the media is not directly established and managed by the public, but indirectly managed by the government that represents the interests of the people. In this situation there can be a misalignment between the will of the government and the will of the people.

The incident of "SARS" in 2002 is a good example. The public was anxious to know the facts of the epidemic, but the media concealed the facts due to reasons such as "to avoid social turmoil. The people became nervous, buying food and medicine excessively. The people, the society, and the media suffered severe losses in credibility. Fortunately, the government promptly corrected this misplacement, and so the media disclosed the epidemic and gave advices to the people to battle against it. Thus, the media regained the trust of the people (Yu et al., 2003b). The threat to the media credibility in China comes from this misalignment between the will of the government and that of the people. One of the ways to reduce this dislocation is to conduct regular surveys of the media credibility so as to understand the public's expectations and assessment of the media.

If the level of the media credibility is determined by the expectations of the public and the actual performance of the media, there are two ways to regulate the media credibility, adjusting the media expectations or a performance. Improving the actual performance and quality of the media is an important focus of journalistic practice and theory. From the perspective of the media credibility, if the media credibility is to be improved, the media performance should be built on the expectations and needs of the audience.

Thus, the media expectation is an important aspect of the media credibility. It is essential to offer media literacy education to the people. Media literacy education can guide the public to establish proper media expectations, but this education is likely to decrease the media credibility in a certain period, such as educating the public to improve their ability to identify "commercial news", which likely leads to a decrease in the trust of the media among those who didn't know the commercial news. In the long run, the improvement of public media literacy can promote the quality of the media, and the media credibility based on rationality is obviously more reliable than the media credibility based on "the foolish people".

Briefly, a summary of the meaning of credibility can be made:

The essential meaning of media credibility is that the public perception and evaluation are the result of the public's social expectations of mass media. This public perception and evaluation are a reflection of the media's capacity to gain the public trust. Simply put, it is a rational reflection of the degree of fit between the public's social expectations of the media and the media's actual performance.

Chapter 2
Dimensions of Judgment of Mass Media Credibility

The basic concept of media credibility and its relationships with related concepts have been defined previously. This chapter discusses the judgment of the media trust and how the public determines whether they trust the media. How is this judgment explained in theory? And what are the contents of the specific judgment and what characteristics do they have? Are there any differences in the judgment of different countries or regions, and what are the rules? Does the same judgment have different meanings in different countries or regions?

2.1 Theoretical Framework: A Cognitive Psychology Approach

2.1.1 Multi-dimensional Judgment of Media Credibility

The same is true of trusting one person, and the reasons for trusting different people may be different. Some people may be trusted for their loyalty, some for their relations, and some for their roles, etc. In other words, people's judgment of trust is not unidimensional. Different people have different dimensions of trust judgment and this judgment is multi-dimensional. Trust is actually composed of the different judgment dimensions. In sociology trust is a multi-dimensional concept which has been generally accepted by scholars. Some scholars think that the definition of trust is ambiguous in previous research. Recent research has explicitly taken into trust as a multi-dimensional structure (Mishra, 2003: 358).

Scholars of Communication define media credibility as a concept of a relational category. They explain media credibility from the perspective of the public, unlike the early "attribute theory" from the perspective of the medium itself. Media credibility is simply an attribute that depends on the medium itself. It is a constant quantity.

G. Yu, *Research on the Communication Effects and Mass Media Credibility in China*,
https://doi.org/10.1007/978-981-19-6242-4_2

From the perspective of the public, different people have different standards of trust in the media because public attitudes are constantly changing.

It is agreed that media credibility is a multi-dimensional judgment. As scholars from Taiwan of China put it, the previous credibility research seems to show that the concept of credibility is fixed and is an object independent of the respondents. However, even if respondents use the same words or concepts, they may refer to different meanings. For example, Zhang San and Li Si both believe that television can be trusted and have high credibility. But Zhang San takes it as the news is fast, while Li Si believes that the news can be trusted because of its sound and light. Although they use the same concept, their definitions of credibility are different (Zhong, 1992: 97).

There are differences between the two common measurement methods of media credibility, relative credibility measurement and absolute credibility measurement. The measurement of relative credibility was conducted by the Roper polling organization in 1959 and has been adapted and developed by scholars (Gaziano et al., 1986: 452). If you got conflicting and different information of the same news story from radio, television, magazines, and newspapers, which of the four channels would you be the most inclined to believe—the one on radio, television, magazines or newspapers?

Given the weakness of relative credibility, such as the inability to measure how credible the news media is and to compare media credibility when the reports are consistent, some scholars, like Carter and Greenberg (1965: 31) propose an absolute credibility measurement method. A better approach to measure credibility is to ask the respondents to assess the credibility of each media respectively. They asked the following question separately for radio, television, and newspapers: We would like your opinions on the reliability of (e.g., radio) news. If perfect reliability is 100%, in your opinion, what percentage of the news on (e.g., radio) do you believe (from 0–100%)?

Whether it is the measurement of relative credibility or the measurement of absolute credibility, results on credibility measures are an overall perception of the media. When using these methods to compare media credibility, the high and low scores are single-dimensional. In fact, there are some other differences in this kind of disparity, because media credibility is a multi-dimensional judgment, that is, the public has different judgment dimensions.

2.1.2 Multi-dimensional Psychological Analyses—The Schema Theory

The mechanism of multi-dimensional judgment of media credibility needs to be further discussed. Why is the public judgment of media trust multi-dimensional? This book uses the schema theory in Cognitive Psychology to explain this concept. The analysis in the following sections of this chapter is based on the schema theory.

Whether or not mass media is credible is achieved through public perception of the media. Scholars in this field seem to have reached a consensus. What is perception? Perception (cognition) is the initial awareness of sensory activity. It shows the process of having consciousness and interpretation of surrounding stimuli or events, which indicates the active selection and understanding of materials from the entire intuitive world, providing a basis for information once it is screened through sensory processes, thereby supporting and guiding subsequent monitoring. A gestalt of cognition (perception) relies heavily on previous knowledge, experience and memory (Fiske, 2004: 203). To explore the multi-dimensional credibility of the media, we need to analyze it from the perspective of the public perception. Social judgment, social cognition and personality theory in psychology have a wealth of relevant literature on how people judge and recognize social situations. Only by making good use of these existing theories can the research of credibility be incorporated into the theoretical framework, so that media credibility research has a theoretical value (Lo et al., 1993).

Take a closer look at social cognition. The basic process of social cognition in social psychology is progressive: from social perception to social impression to social judgment. According to the interpretation of scholars in social psychology, social perception is the first step in the whole process of social cognition. It is a holistic perception of various attributes and characteristics of cognitive objects themselves. Social impression is a social psychological phenomenon that is formed on the basis of social perception. It is the image of the cognitive object that people form in the mind and stay in memory through contact and perception with cognitive objects. Social judgment is the evaluation and inference of cognitive objects based on social perception and social impression (Zhou, 1997: 170–176). It can be seen that the perception of media credibility is finally in the form of judgment.

But the basic process of social cognition discussed above does not explain the question of "why the public judgment of media credibility is multi-dimensional." How do the public form their perception, impression and final judgment about the trust of the media? What is its internal mechanism? It is supposed to use the cognitive structural theory in cognitive psychology. This argument was raised by communication scholars in Taiwan, China more than a decade ago. In the past decades the cognitive structural theory was dominated in the study of social cognition. We believe that the credibility research is a kind of social cognition, and it should be explained by using cognitive structural theory (Zhong, 1992: 98–99). Specifically, it is the cognitive schema.

From the perspective of epistemology, Piaget, a Swiss psychologist, thinks that soon after the individual was born, they began to use some of their inherent behavioral patterns to respond to their environment. It can be regarded as the cognitive structure for an individual to know the surrounding world. When an individual encounter something and processes it with a corresponding cognitive structure, this cognitive structure is called a "schema". According to Piaget, the schema is a basic framework for individuals to absorb new information. It is the basic unit of cognition. This framework changes with age or knowledge.

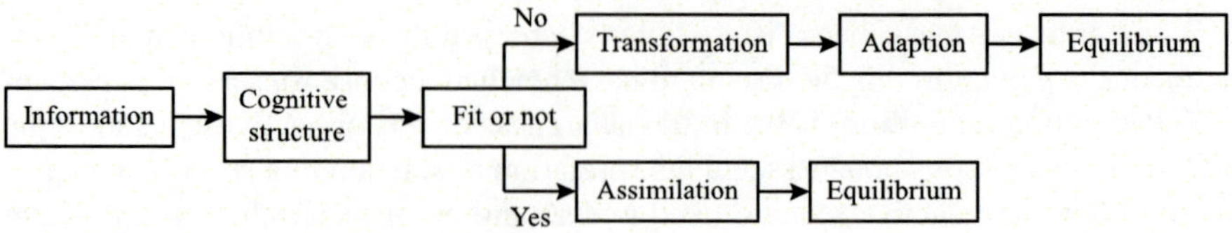

Fig. 2.1 The schema mechanism of cognitive structure

Furthermore, Rumelhart defined schema as "a data structure for representing the genetic concepts stored in memory". A set of schema constitutes schemata. In short, the schema theory explains how these schemata are represented, and how these representations are used in special situations. The Chinese scholar Lin (2004: 146) illustrates the schema of cognitive structure as shown in Fig. 2.1.

Lin (2004) thinks that both Journalism and Communication use the concept of schema to explain that people use schemas to identify, understand and manage information which represent objective things. The recipient does not know the meaning of unfamiliar information in the psychological system when they have access to it for the first time. The psychological system puts this information into various schema, and when this information matches a certain schema, the system judges that the things this information refers to is the thing the schema represents. Therefore, the schema is used to refer to this information. Tens of thousands of schemas existing in brains of adults are not inherent, nor are they produced in a sudden, but are gradually accumulated from infants to adulthood.

For receiver's cognitive schemas in mass communication, some scholars sum up three characteristics. First, it is subjective, which refers that different cognitive schemas are the same or similar media information produced in brains of different receivers. Second, it is hierarchical, which means that the network of schemas are progressive formed from low to high according to the content of its representation in the cognitive structure. Third, it is mouldable, which means that a schema is relatively stable once the schema is formed. However, to adapt to the changing world, recipients must enlarge or deepen existing schemas (Liu, 1997: 136–138).

Thus, the cognitive schema theory is employed here to examine the multi-dimensional feature of judgment of media credibility. At the individual level, schemas of individuals are different. For different schemas have different perceptions, impressions and judgments for the same medium, so media credibility may have different judgment dimensions. At the regional or group level, people in the same region or in the same group establish a similar schema structure since the environment is similar. The judgment of media trust has a certain similarity. For different regions and groups, due to their different schema, their judgment dimensions can be different.

In the study of mass media credibility, the book does not explore the specific mechanism of the cognitive schema but employs this theory to analyze the trust relationship between the public and the media, including characteristics of credibility, variability of its constituents and differences.

2.2 Composition and Characteristics of Dimensions of Judgment

2.2.1 Composition of Dimensions

As discussed in the previous chapter, media credibility is not a unidimensional, but a multi-dimensional judgment. The following is to examine the composition of dimensions of media credibility judgment. The dimensions of trust are first discussed.

2.2.1.1 Multi-dimensional Trust

As for dimensions of trust judgment, some representative research are conducted by Aneil K. Mishra.

From the perspective of organizations Mishra (2003) studies the trust relationship among individuals, organizations, and between organizations and individuals in a firm, and argues that trust is one party's willingness to be vulnerable to another party based on the belief that the latter party is competent, open, interested and reliable, that is, trust has four dimensions: competence, openness, interest, and reliability.

(1) *The competence dimension of trust*

Mishra thinks that within organizations, managers develop relationships with their subordinates and with other managers largely on the basis of trust which is defined in terms of competence. At the organizational and interorganizational levels, the competence dimension of trust is also discussed, especially in the context of exchange relations. For example, if the product of a supplier meets quality standards of the buyer, after dealings for a while, the buyer no longer inspects products before accepting delivery, which shows that the supplier has the competence to win the buyer's trust.

According to his interviews with managers, Mishra finds that competence is a dimension of trust that exists between individuals and organizations. He discusses with a manager about how to maintain and develop trust in the organization, the manager stated:

> They've got to have some feeling that you're competent to lead them out of this mess. Because, they may like you a lot but if they feel you're a bumbling idiot they say, "We can't trust what this guy tells us. He's gonna take us off the end of the cliff." I mean they have to be confident that you're competent. They've got to have some feeling that you know what the hell you're talking about. When you go out there to tell them to do something they've got to have some feeling that it will make a difference.

(2) *The openness dimension of trust*

Scholars have done some discussions and empirical research on the openness of trust in organizations. Davis and Lawrence (1977) defines the concept of trust, and its openness which is able to coordinate trust in matrixed departments. In the research of trust construction, Butler (1991) finds that openness serves to establish trust.

Openness is a dimension of trust, but Mishra further finds that this dimension is a double-edged sword. Openness beyond a certain level may, however, serve to impair rather than enhance trust. For example, telling someone the complete truth with elaborate details about his or her character flaws may decrease trust between two parties. Nevertheless, such extreme honesty impairs the overall trust by lowering trust in dimensions of competence and interest, rather than the openness dimension.

(3) *The interest dimension of trust*

Mishra argues that the interest dimension of trust means that one party believes that it will not be taken unfair advantage of by the other party. For example, I can believe that another party not only refrain from taking unfair advantage of me, but this party is also concerned about my interests or the interests of the whole. This does not mean that the other party lacks any self-interest. Rather, trust in terms of interest means that such self-interest is balanced by interest in the welfare of others.

Through interviewing with managers Mishra proved this dimension:

> I think the trust factor is that you have the best interest of the people at heart, and that they believe in what you're trying to do...If they feel that you really care. I have a responsibility for 3,000-plus people and their families. That's an awesome responsibility. My single objective is to make sure that those people have a livelihood, a future, so that they can take care of those families. That's all part of this trust issue.

(4) *The reliability dimension of trust*

Reliability refers to consistent or reliable behavior and the consistency between words and action. Reliability is a dimension of trust that has been discussed by many scholars. McGregor (1967: 164) argues, inconsistencies between words and action decrease trust. Ouchi (1981: 101) referred to trust in terms of expectations about consistent or reliable behavior. Trust between managers and subordinates is also defined by Gabarro (1987: 105) in terms of consistency of behavior.

In Mishra's interview on trust, reliability is repeatedly raised by interviewees as an important component of trust.

> I'd say for the most part, our level of trust with our customers is pretty good. Because we must do what we say. If we say we're going to have it on the 17th, we have it on the 17th. If something would prevent that from happening, we warn them up-front. We work with them on an early warning system. There again, customers only trust you because you have a history of delivering what you say.

Another manager from a supplier firm explicitly compares customer firms in terms of their reliability:

> In most cases, we've certainly developed more of a trusting relationship with certain companies than we've been able to with this other company. I believe it's because their middle managers are getting consistent messages from the top. These companies are less likely to pull the rug out from under the people we deal with directly. Whereas, the other company's (top managers) will change their minds overnight, and all the things that are put forth by the people that I deal with on a one-on-one basis suddenly aren't valid anymore.

2.2.1.2 The Multi-dimensions of Media Credibility

Four dimensions of trust (competence, openness, interest and reliability) also exist in media credibility with the same basic meanings. However, unlike the subject and object of trust are between individuals, organizations and individuals and organizations in an enterprise, the subject and object of media credibility are the public and the media. The extensions of media credibility include the credibility of media channels and institutions, etc. Therefore, four dimensions of trust appear in different extensions and are different from the dimensional structure of each extension of media credibility.

In the field of communication academia, scholars examine the dimension of credibility judgments following two extensions of media (channels) and sources. Kiousis, a communication scholar, claims that it is important to note that some empirical work has indicated that it is meaningful to distinguish between source credibility and medium credibility, although two terms obviously overlap. Thorson and Moore (1996) asserted that credibility is an issue for both *source* as well as for the *channel* or *medium* carrying the message. As a result, it appears sage to draw on literature from both research approaches to avoid theoretical and empirical confusion (Kiousis, 2001). In addition to the message source and the media (channels), some scholars explore the connotation of credibility from the perspective of a media institution. This book discusses the compositions of these dimensions including the credibility of the message sources, of the media (channels) and of the institution.

From the perspective of the schema theory, the public has different cognitive schemas for the media sources, channels, and institutions, which together constitute the schemata as a whole. In other words, first, there is a cognitive schema for media channels from the reception of news. For media channels such as newspapers, televisions, and broadcasts, there are different functional judgments and different levels of trust. This affects the audience's choice of channels. Second, there is a cognitive schema for the institutions. For different institutions, the audience have different trust judgments since they choose different institutions based on different needs for news information. For example, in the coverage of the Afghanistan war, some people trust CNN, some people trust BBC, some people trust Xinhua News Agency, and other people believe in Al Jazeera, and so on.

Third, there is a cognitive schema for message sources and contents. If the news is provided by Xinhua News Agency or by a citizen, there is a difference in credibility. Also, for the same channel, different sections and different hosts can create different perceptions.

Let's look at the empirical research findings of American scholars to examine three dimensional structures of media credibility. Empirical research of communication is mainly conducted by American scholars, and much research on media credibility is done in the United States. Chinese scholars have just started research on media credibility, and there is no empirical research data there.

The credibility of the source mainly explores the trust level of the communicators (including individuals, groups, and organizations). Hovland, Janis and Kelley (1953) focuses on the message source in their early research. It empirically identifies expertise and trustworthiness as the main dimensions of the credibility of communicators or sources. Markham (1968: 57–64) takes journalists as the research subject, and has undergraduate students view kinescopes of three local newscasts and evaluate each of the newscasters on a fifty-five item semantic differential instrument and identifies three major dimensions of source credibility as a reliable-logical or validity of the message factor; a showmanship, dynamism, or entertainment factor; and a trustworthiness dimension. Through a series of factor analyses of the semantic differential scale, Berlo et al. (1969) claim that source credibility has three dimensions: safety, qualification, and dynamism. After the 1990s scholars broadened the scope of research on the credibility of message sources. Slater and Rouner (1996) argues that internal characteristics of messages, such as their aesthetic presentation or their actual content, can sway perceptions of source credibility.

Meyer's research involves the dimension of the channel credibility judgment, and finds that no matter whether the credibility of newspapers or television news, there exists two factors, i.e. believability and community affiliation. It mentions the reliability and validity of each measurement item that constitutes the dimension of "trust" and states that fairness, unbiased, telling the whole story, accuracy and trustworthiness are the best indicators (Wang et al., 1999).

On the credibility judgment of the press, Times Mirror (1986) finds that the credibility of the press consists of four dimensions: the special interests and the press, the power structure and the press, the press and its performance, and the character of the press (Lo et al., 2003).

2.2.2 Characteristics of Credibility Judgments

Regarding dimensions of media credibility judgments, this book summarizes three characteristics which are integrity, relevance, and variability.

2.2.2.1 Integrity

Since media credibility is defined as a multi-dimensional concept, what is the relationship between these dimensions?

The relationship between dimensions of trust is first analyzed. Through the empirical research, Aneil K. Mishra verifies four distinct dimensions of trust, and further analyzes the relationship among these four dimensions. In his view each dimension of trust is assumed to represent components of the overall trust structure. These dimensions combine multiplicatively in determining the overall degree of trust that one party has with respect to a given referent. That is, a low level of trust in terms of any of dimensions offsets high levels of trust in terms of the other dimensions. For example, a manufacturer could trust one of its suppliers to be competent in producing particular engine components and to be reliable in meeting its delivery schedule. But the manufacturer might still suspect that the supplier does not report its true cost estimates in order to price-gouge when selling those components to the manufacturer. In other words, the manufacturer may not trust the supplier in terms of openness or interest. The manufacturer in this instance would then be expected to have a low level trust in that supplier.

Similarly, various dimensions of media credibility judgments are in a holistic relationship. The judgment of whether people in the same region or the same group trust the media is multi-dimensional. The connotation of media credibility is composed of these different dimensions. The decrease of each dimension may impair the overall credibility of the media.

According to this perception, in addition to the relative and absolute measurement methods introduced above, scholars have developed "multi-angle measurement methods" to measure media credibility and to survey various dimensions that constitute media credibility. The sum of the survey results is media credibility. As a result, the decrease of a certain dimension affects the credibility of the entire media.

2.2.2.2 Relevance

We have discussed the differences in dimensions of judging the credibility of message sources, media channels and institutions. One of the questions we need to answer is whether the relationship between dimensions of their credibility is completely separate.

Kiousis (2001) thinks that the message source and the channel are not completely separate in practice, but they are not exactly the same, and states that boundaries between channels and source credibility are somewhat blurred but are associated because of their common conceptual foundation. This explication is more expansive than prior models. So they do not distinguish between the terms, neither do they share common theoretical framework, because it allows for differences in concepts but admits that they do share common qualities. As such, this project essentially measures opinions about media channels, but perceptions of sources (e.g., journalists and news organizations) are also inevitably impacted by such opinions. People's impressions of channel credibility may drive their opinions about source credibility, but in turn, opinions about source credibility may drive impressions of channel credibility.

Credibility of media institutions, same as channel credibility and source credibility, is conceptually distinguished, but their conceptual foundations are associated, and there is no clear boundary. The institution credibility is built on sources and channels. Therefore, these three can be distinguished conceptually, but the nature of public evaluation is the same.

The public has separate cognitive schemas for each of them, and their credibility judgments have different dimensional indicators. They are interrelated and mutually influenced. For example, to provide vivid and realistic content is an important way for the TV to win public trust, but this feature is shared by channels and institutions. The quality of the communication content directly affects the trust of channels.

2.2.2.3 Variability

Scholars of the empirical school of communications have done much research on dimensions of media credibility judgment. So far there has been no widely accepted dimension scales, and dimensional structures obtained by different studies are different. This is due to the different design of each study, and these studies were conducted at different times and places. Thus this dimension of media credibility judgment is not fixed and is variable.

From the perspective of the schema theory of cognitive psychology, the schemas of people are different when surveyed in different groups at different times and places. Even for the same research design and the same population, the dimensional structure obtained at different times may be different. People's cognitive schema are changing constantly. The greater the difference between time and region, the greater the difference in dimensions of judgment derived from the study. In short, the composition of media credibility changes with different times and places.

2.3 Comparison of Regional Differences in Dimensions of Judgement

The dimension of media credibility judgment has characteristics of variability, since people affected by different environments may build different cognitive schemas and have different dimensions of judgment. At the national or large regional level the environmental differences are political and cultural differences. These differences may affect people's cognitive schemas of media credibility judgment and form different dimensions of media credibility judgment. We select the United States for comparison with China.

First, we examine the dimensional composition of the United States. Then the literature review and in-depth interview are analyzed to explore the composition of the judgment dimension in China's mainland and the Taiwan region; second, we compare the composition of dimensions of media credibility in China and the United States, and examine the influences of two macro variables of politics and culture on the composition of dimensions of credibility judgment. Finally, we analyze reasons for regional differences in the composition of dimensions.

2.3.1 *Composition of Dimensions of the Subject*

2.3.1.1 Composition of Dimensions of Media Credibility Judgment in the United States

For the dimension of the credibility judgment of the American media, we have discussed dimensions of credibility, i.e. the media channels, sources, and institutions respectively. Here is a brief introduction to the findings of American scholars. Hovland et al. conducted research on media sources in the early 1950s and found that credibility had two important dimensions which were expertise and trustworthiness. Other scholars continue to do empirical research on the media channels sources, and institutions.

In 1966 Jacobson used 20 semantic-differential items in his dissertation to perform a factor analysis and proposed four dimensions of source credibility, which were authenticity, objectivity, dynamism and respite. Authenticity included trustworthy-untrustworthy, accurate-inaccurate, expert-ignorant and openminded–closeminded. Objectivity included unbiased-biased and impersonal-personal. Dynamism included colorful-colorless, important-unimportant, easy-difficult and bold-timid. Respite included relaxed-tense and pleasant-unpleasant. Jakobson (1969) argues that authenticity and objectivity are two credibility dimensions of credibility while dynamism and respite are two non-credibility dimensions.

In his research in 1976 Singletary found that media credibility was presented through six dimensions, including knowledgeability, attraction, trustworthiness, articulation, and hostility and stability. In 1978 Lee made a further study and found that the dimensions of media credibility were affected by different media and the nature of news. It was also found that when respondents were asked to evaluate national and international news in the newspaper, there appeared four dimensions of credibility including trustworthiness, intimacy, expertise and availability. Meanwhile, when they were asked to evaluate national and international news on TV, there appeared three dimensions of credibility including bias, intimacy and dynamism. When evaluating local and state news in the newspaper, there appeared four dimensions of credibility, which were trustworthiness-authenticity-agreeableness, dynamism, intimacy and bias. When evaluating local and state news on TV, there appeared three dimensions of credibility, which were reliable-real, immediate- intimacy, and vitality-expertise (Lo et al., 2003).

Gaziano and McGrath conducted a research on credibility of American newspapers in 1986, which attracted much attention. They (1986: 451–62) used a set of scales with sixteen bipolar semantic differential items developed by the American Society of Newspaper Editors (ASNE). The result of the factor analysis showed that a dominant factor was "credibility", including trustworthiness, accuracy, fairness, respect privacy, etc. Another factor is social concerns, including being immoral, not caring about what the reader thinks, and sensationalizing.

Meyer argued in 1988 that the positive and reverse-polarity items used by Gaziano et al. were mixed and might mislead respondents, and they lacked reliability and theoretical basis. As discussed in the previous dimensional structure, Meyer found that there were five dimensions: "fare", "unbiased", "tells the whole story", "accurate" and "can be trusted" (Wang, 1999).

West, Meyer's colleague, reanalyzed the data collected by Meyer in his study in 1994 and agreed that the "credibility" consists of these five items. Compared with the "credibility" composed of twelve items, it was more effective in measuring the credibility of newspapers and television news. From reanalysis of the measurement results of other items it was difficult to find that "social concerns" or "community affiliation" is a separate factor in media credibility (Wang et al., 1999).

In 2000 Flanagan et al. argued that media credibility had been discussed for more than 50 years, and these studies showed that different research methods lead to different concept categories. The most consistent dimension of credibility is believability, but accuracy, trustworthiness, bias and completeness are other dimensions often used in research. Thus, media credibility is a multi-dimensional concept. For the media some scholars have done much work based on the Gaziano-McGrath studies and made further studies. As noted above, media credibility in the United States has five dimensions: "fare", "unbiased", "tells the whole story", "accurate" and "can be trusted".

Regarding the dimension of the source credibility, although scholars have no consensus on its key meaning, most scholars agree that expertise and trustworthiness of source are two main dimensions of credibility (Lo et al., 2003; Hovland et al., 1953; Kiousis, 2001).

2.3.1.2 Composition of Dimensions of Media Credibility Judgment in China

(1) *The composition of media credibility in China's mainland*

As there is no empirical research data in this area, this book discusses dimensions of media credibility in China's mainland from two aspects. The first is the literature review which examines the credibility discussed by scholars and reported by news reporters. The second is empirical research conducting in-depth interviews with 30 samples (Lin et al., 2003: 21). Based on the findings the composition of media credibility in China's mainland is examined.

The following is written by scholars.

> According to relevant surveys, "advertisements with a high proportion of false elements in China's mainland have a trust of more than 60%". After careful consideration, it is not difficult to find the reason that our media is official. The recipients have a trust of 60% in advertisements, which is actually the trust of "official" media and the trust of government authorities. The media plays a "propaganda" role, as a megaphone for business, out of which there are some false advertisements. In this way, it is easy to mislead the receiver perception of consumption and even culture and values. Therefore, the great danger lies in the blind trust of receivers in the media authority.

Generally speaking, things with authority can gain the trust of the people. The Party newspaper is a press that the Party and the government rely heavily on. It is a news media with high authority and influence among the people. Marx emphasized, "the trust of the people is the condition for newspapers to live. Without this condition, the newspapers will shrivel". Economic report, as an important content of the Party newspaper, depends on the trust of the people (Ding et al., 2004).

The above states that an important element of media's capacity to gain the public trust is the authority of the media, since the media is "mouthpiece" of the Party. The first paragraph analyzes one of the reasons why so many false advertisements in media with such a high degree of trust is the official nature of the media. In other words, the public trusts media advertising based on the trust of government authority. The second paragraph shows that the Party newspaper gains trust because of its authority.

The above analysis is based on the interview with Liu Wenbao, a farmer from Hubei, by *Southern Weekly*. Liu Wenbao loved the work of turning the deserted beach into a grain field, and the state leaders praised him as a "king of grain". However, he was subjected to various difficulties. He moved several times and was in a lawsuit. Being interviewed by reporters, he enjoyed "the attention made by the news media, which is more effective than the intervention of leaders" (Zhu, 2005).

Next, based on empirical in-depth interviews media credibility is examined from the perspective of the audience.

From November 13 to 20, 2004, 30 sample interviews were conducted among readers of *Beijing Newsstand*. There were three questions in the interview outline:

① The newspaper you bought today is: ____________________

② For the newspaper you bought, do you believe in the various news on it? If 100 points refer to the complete trust and 0 is not to trust completely, what do you think the scores of various types of news are:

Domestic affairs news

International news:

Local social news:

Sports news:

Entertainment news:

Finance and securities news:

Advertisements:

③ Why do you think so: ______________

The survey showed the interviewed readers bought *Beijing Evening News*, *Beijing News*, *Beijing Times*, *Beijing Daily Messenger*, *Beijing Youth Daily* and *Southern Week*ly. Domestic affairs news, international news, local social news, finance news, and securities news all got high scores, generally over 80 while entertainment news scored between 50 and 70 points, and advertisements received the lowest score.

Why did readers trust the news in the newspaper? Nearly half of them said they could not tell. After questioning, they expressed reasons why they trusted these newspapers. First, a common statement is that newspapers are run by the government; they are authoritative and should be trusted. "Official", "authoritative" and "governmental" are words frequently used by interviewees. Nearly one-third of the interviewees stated these reasons in the first response. Other expressed the same idea in their conversation. Second, some interviewees thought that the media was useful. For example, corruption and unfair things was reported in the media. In particular, many people said that they would complain to the media when they encountered consumption problems. Third, there were requirements for the authenticity of news. For example, many people thought that the entertainment news was made by paparazzi where there was more hype and sensationalism. Advertisement exaggerated facts, and probably provided misinformation. The judgment criteria for a sports game were the same, and there was no falsehood. There were some interviewees believing that the truth of the news would affect their trust in specific media. For example, an insurance saleswoman of American International Assurance thought that *Beijing News* had 80% credibility and *Beijing Daily Messenger* had only 30% credibility, since it reported on a matter related to her company. She found that the report was inconsistent with the fact, so she didn't trust *Beijing Daily Messenger* anymore.

It's not enough to draw an overall conclusion from interviews of these 30 samples. But in the absence of more empirical data the interviews could reflect basic dimensions of media credibility in China's mainland. The main point is that these in-depth interviews explore the people's deep and psychological perception of trust, which is a common method of exploratory research. With analysis of other scholars, we can at least make a preliminary conclusion that one of the primary dimensions of the public judgment of media credibility in China's mainland is "authority", which means "government" and "government-run". "Practicability" is an important dimension. Liu Wenbao argues that newspapers could solve problems or act as a channel for the people to make complaints. In addition, "authenticity" is a dimension by which the public judge the credibility of the media.

(2) *The composition of media credibility in the Taiwan region*

The research conducted by scholars in the Taiwan region uses the empirical methods and theoretical models of American scholars.

Lo Ven-hwei et al. conducted research on the index of credibility in the Taiwan region, based on the scale used by ASNE (1985) and modified according to the results of focus group interviews on newspapers and televisions with sixteen items respectively. According to ASNE's research these sixteen items constitute two dimensions, "credibility" and "social concerns". Sixteen items are shown in below table Table 2.1.

The principal factor analysis of the survey data shows that the above sixteen items do not form "credibility" and "social concern" as the ASNE study did. There is one factor in the evaluation of newspapers and television. The sixteen items such as the reliability and accurateness along with the public interest and the national interest

Table 2.1 Two dimensions of judging the media credibility and sixteen items

Items for credibility	Items for social concerns
Reliable	Concerned about public interest
Accurate	Value national interests
Objective	Concerned about social well-being
Fair	Respect public opinion
Comprehensive	Respect people's privacy
Unbiased	Uphold the right of the public
Not exaggerated	Promote social reform
Trustworthy	Uphold social justice

constitute one factor, which together produces the index of the public judgment of media credibility (Lo et al., 1993).

Zhong (1992) did a survey on the judgment of media credibility from the perspective of the ethnomethodology theory. With open-ended questions respondents were asked to provide reasons after they made an overall evaluation of the media (believed or unbelieved). About 77.2% of the respondents clarified the criteria they used to evaluate the media, as shown in below Table 2.2.

Lo Ven-hwei et al. used ASNE's scale in their research. Although there were previous interviews and some modifications, a framework was set up at the start of the research. The discussion was conducted within the framework designed by researchers in advance. Otherwise it was impossible to ensure that the ideas of the public in the Taiwan region would be included, and to find the principal index of the credibility judgment in the Taiwan region.

Zhong Weiwen used open-ended questions to perceive the public judgment. According to his research, the public in the Taiwan region do not have such complicated ideas as the sixteen items listed by Lo Ven-hwei, and only 5% of the public have more than one standard.

Thus, his research better displays the public judgments of credibility in the Taiwan region. The major indices are accurateness, impartialness, professionalism, and independent stance.

Table 2.2 The criteria to evaluate the media credibility

Accurateness	34% (n = 77)
Impartialness	12.3% (n = 28)
Professionalism	10.6% (n = 24)
Independent stance	6% (n = 14)
Other	19.4% (n = 44)
Blurred	17.2% (n = 39)
Total number of samples	n = 226

Table 2.3 Dimensional indicators of the credibility in three regions

The United States	China	
	The mainland	Taiwan
Accuracy	Authority	Accurateness
Fairness	Practicality	Impartialness
Expertise	Authenticity	Professionalism
Unbiased		Independent stance
Completeness		
Reliability		

2.3.2 *Comparison of Results and Influencing Factors*

The judgment dimensions of media credibility in China, and the United States are respectively discussed above. Now similarities and differences of the judgment dimensions of media credibility in China and the United States are analyzed. The first is a comparative analysis of the similarities and differences in the judgment dimensions of media credibility; then, the triangulation approach proposed by Professor Zhu Jianhua of the City University of Hong Kong is used to examine the impact of the social macro variables, which are politics and culture.

2.3.2.1 Composition of Judgment Dimensions

Based on the previous research on the composition of the credibility in China and the United States, dimensional compositions are comprehensively compared. Whether it is a message source or a media channel in the United States, dimensions of credibility basically cover fairness, unbiasedness, completeness, accuracy, and expertise. In Taiwan of China, according to Zhong Weiwen's research they are accurateness, impartialness, professionalism, and independent stance. They are authority, practicability and authenticity in China's mainland. The dimensional indicators are shown as Table 2.3.

2.3.2.2 Political Determinants Behind Similarities and Differences in Dimensions

Zhu Jianhua et al. from the City University of Hong Kong made a comparative study of journalists' perception of the media's role in China and the United States, and found that social factors were more important than the latter two in terms of social, organizational, and personal factors. Personal factors have little effect. Social factors account for 50–74% of the cognitive variance of media's role, while the organizational factor of the media industry only explains 3–4%. The personal factor in professional training can only explain 1%, or even less than 1% (Zhu, 2003: 467).

The public perception of media credibility in China and the United States is explored at the macro social level. The analysis is based on the triangulation approach proposed by Professor Zhu Jianhua.

First, the analytical framework of triangulation approach is shown in below Table 2.4.

This framework is tested and explained as follows:

(1) Determinism of variable one (specifically represented by case $1 = 3$, $1 \neq 2$, $2 \neq 3$);
(2) Determinism of variable two (specifically represented as case $1 = 2$, $1 \neq 3$, $2 \neq 3$);
(3) The identity of the mastermind is unknown (specifically, case $1 \neq 2 \neq 3$, because both variables have no effect, indicating that this study failed to find macro variables that have a real impact);
(4) "Social impact does not exist" (specifically, case $1 = 2 = 3$, indicating that not only the two variables being compared but also any other variables not being compared have no impact);
(5) Theoretically impossible phenomenon (i.e. $2 = 3$, $1 \neq 2$, $1 \neq 3$, because 2 and 3 have different values on the two variables, this is theoretically impossible. If this anomaly occurs, it often implies that there is a problem with data collection and analysis).

Second, for the above comparison framework, Zhu (2003) converted to a specific analytical model when conducting research on journalists' perception of the media's role in China and the United States. Using this model on a social level, the role of "political determinism" and "cultural determinism" was compared, and the result is consistent with the model predicted by "political determinism", that is politics has a greater influence than culture. The specific model prediction is shown in below Table 2.5.

Finally, the two macro variables of politics and culture are compared with reference to the above analysis model. For the application of the triangulation approach, Zhu Jianhua uses data for quantitative analysis. Due to the lack of basic data, this book borrows the idea of the triangulation approach and uses the conclusion of credibility judgment dimension for comparison. It's just a qualitative discussion.

According to the previous comparative analysis of the credibility dimensions, political factors have a decisive effect on the composition of the credibility.

Table 2.4 The analysis framework of triangulation approach

Item	Variable two	
Variable one	Value A	Value B
Value A	Case 1	Case 2
Value B	Case 3	(Case 4)

Table 2.5 The model prediction applied in China and the United States

Model prediction	The United States	China	
		The mainland	Taiwan
Political determinism	Same	Different	Same
Cultural determinism	Different	Same	Same
Irrational concept	Same	Same	Different
Wrong designation	Different	Different	Different
No social effect	Same	Same	Same

2.3.3 Two Dimensional Orientations with Different Political Factors

The previous comparisons show that the composition of media credibility in China and the United States is dominated by political factors at the macro-social level. This section explores the relationship between political factors and media credibility, and the characteristics of dimensional composition of media credibility. The first is to review the research on types of government-media relations. Second, it analyzes the media functions in China and the United States under two different political systems. Finally, characteristics of media credibility under the influence of different political systems are summarized.

2.3.3.1 Type of Relationship Between the Media and the Government

In 1956 Frederick Siebert, Theodore Peterson and Wilbur Schramm co-authored *Four Theories of the Press*, which made a wide-ranging impact in academic circles. They proposed four theories, i.e. the Authoritarian Theory, the Libertarian Theory, Social-Responsibility Theory and Soviet-Communist Theory.

In the 1980s four theories on the press began to encounter big academic challenges. J. Herbert Altschull argues in the book *Agents of Power* that the "four types of press theories" don't work anymore to explain the current media system. He points out that the "independent press" no longer exists and in a social system the news media are agents of those who exercise political and economic power.

Peng Huaien from Taiwan of China provide a new explanation of the four theories of the press and propose four types of relations between the media and the government with the development of information technology.

(1) *The relationship between the media and the government in the authoritarian press system*

The relationship between the authoritarian media and the government is the top-down control by the government. The media is fully regulated, and there is no "press freedom" at all. The media is trying to get out of the government's control but often encounter setbacks and restrictions. The government sometimes wants to use the media as a tool of governance, but it is a subsidiary with the purpose of controlling. Many of its measures often stem from a preventive mentality. Sometimes the government or the media is mobilized to engage in propaganda. The media is only obedient, and their activities cannot compete with the government's political action.

This kind of relationship between the media and the government exists in the agricultural society or in the early and middle industrial society. By the mid-term industrial society, the middle class has grown with social prosperity, the advancement of civilization, the concept of democracy and freedom. The full control of the government cannot continue to be so effective, and the relationship between the media and the government is changing.

(2) *The relationship between the media and the government in the libertarian press system*

With the rise of the democratic revolution of the middle class, more freedom of the press is demanded. The relationship between the media and the government has come to be equal and independent. The relationship can be divided into two types:

The media and the government would be parallel. They check and balance each other. This is the current concept of the press held by many British and American journalists who believe that this is in line with the spirit of free press. The media itself has considerable autonomy. Although the government is governing the country based on the consent of the people, the media monitor the government as a representative of the people. The media is the fourth power (or the fourth sector) to check the executive, legislative and judicial powers. It checks and balances the government, provides sufficient information for the people, and acts as a protector of social welfare.

The media and the government cooperate with each other. They believe that the press has its own autonomy, but the media should not be opposed to the government. They should be good friends and work together. Their relationship is interactive and basically harmonious. The media should work with the government to perform their functions. The best way is to serve as a bridge between the people and the government. However, this kind of relationship has its dangers. Sometimes the media is manipulated by the government. Especially since the government controls many resources, reporters are easily bought over by government officials and lose their independence.

(3) *The relationship between the media and the government in the communist press system*

In the communist press system the relationship between the media and the government is that the media is under the full control of politics (the government or the

Party's department). It is a department of government. In this context the press system is theoretically and practically part of the political system, and it serves the government and the people. News communication activities are only a part of political actions, and media activities are political ones, but political actions are not necessarily subject to check and balance by the media.

(4) *The relationship between the media and the government in the information society*

After the society enters into the information age or the Internet world, the whole social life is changing greatly, and the relationship between the media and the government is changeable.

In the information society media activities are ubiquitous and dominant in social life. The population engaged in information work exceed the industrial population, but in the case of information as power, the administrative authority of the government is relatively reduced. Especially in the computer utopia, due to the openness of society and the maturity of human beings the relationship between social dynamics and checks and balances naturally operates in the flow of information. Therefore, the government and its laws and regulations may be prepared but not used. Political activities may be a small part of social activities. Instead, the government yields to the reality of information and lose some of its functions. At this time there will be many voluntary groups to replace some functions of the government.

2.3.3.2 Functions of the Media in Different Political Structures

Four types of government-media relations proposed by scholars from Taiwan of China have developed the theories of Schramm and others, especially focusing on the relationship between the media and the government with the development of new technology. Judging from the four types of relationships between the media and the government discussed above, in the United States there is the second type of "media-government relationship in the libertarian press system". In China there is the third type of "media-government relationship in the communist press system". However, the political control under this American libertarian press system is insufficiently analyzed. Actually the media don't have real freedom, and they are manipulated by political and economic forces. The fourth type of "relationship between the media and the government in the information society" is not yet a reality, but a possible future for the development of information technology. However, the information society is a trend, and the relationship between the media and the government in the information society is theoretically forward-looking.

The following is a detailed analysis of the relationship between the government and the media in the United States and China and the functions of the media are examined.

The political structure of the United States is a federal system. The form of political power is a presidential system. It implements a political system that combines separation of powers and checks and balances together with a two-party system. The organization of the regime adopts a presidential system, and the president is the head of the state and government. The decentralization and checks and balances are implemented. Legislative, executive, and judicial powers are controlled by the Congress, the President, and the Court. When the three branches exercise power, they check each other. China is a multi-party cooperation and political consultation system under the one-party rule of the Communist Party. Thus, the political structure of American society is horizontal while that of China is vertical.

In a horizontal social structure, the mass media are private institutions ostensibly independent of the government (except for a few media such as public television). To a certain extent they are supervisory forces that are parallel to political power. They are the fourth power (or fourth sector) to monitor over the power of the executive, legislative and judicial branches. There is a certain amount of independence in checks and balances of these powers. This independence of the media is relative, and is often manipulated by political and economic forces. In such a political structure the media and the government display an independent and separate relation. It is difficult to fully understand the manipulation of political and economic forces behind media independence. Thus, the public expectation for the media is mainly the function of "watchdog" and serve as a lookout post for society and a supervisor of power. According to a survey in the mid-1980s, 50–70% of Americans agreed with the role of "watchdogs" in the press, and that they watch the behavior of political leaders and prevent them from doing what they should not.

In the vertical political structure of China, the media is the "mouthpiece" of the Party, an extension of government functions and a propaganda machine for the Party and the government. The Party and the government control the media and the media is subordinate to the political power system. The relationship between the media and the government is integrated. Chen (1999a), a scholar at the University of Macao in China, believes that the role of government's "mouthpiece" gives the media a huge "asset." First, the media has a close relationship with the Party and the government and enjoys the prestige and reputation that ordinary enterprises cannot reach. Second, the government provides financial "security" and benefits to the media. Finally, the government maintains a "monopoly" for the media. The media directors are actually agents of the government (or the superior department) in the press. Their responsibility is to ensure the normal operation of the media and provide publicity and public relations services for the government. The government provides adequate financial security to the media and its employees and offers media executives an opportunity to advance within the administrative system.

For the people in China, the media is more of a government spokesperson, and their expectation for the media is more of their role in the government.

2.3.3.3 Characteristics in Two Political Systems: Professional Orientation and Power Orientation

In the horizontal political structure of the United States where the media is relatively independent of the government, it is one of many social sectors. The public expectation for the media is more of a role of watchtower. It is expected that the media can monitor power. People's trust in the media is more about their professional performance. For the media to play this role well and gain the public trust, they must try to meet the professional standards of accurateness, reliability and professionalism of news reports.

What is professionalism in journalism? According to Zhan (2004), a scholar from the Department of Journalism of the China Youth University of Political Studies, journalism is an important concept in western journalism. The goal is to serve the public rather than a certain interest group. In media institutions with dual natures of commerce and politics, they emphasize that they are the defenders of the public interest so as to gain public trust. Zhan Jiang believes that professionalism in journalism has two main characteristics, i.e. a neutral gatekeeper and an objective reflector. In the business world the media is independent of the government and the market and uses professional values and forces to strengthen the journalist's psychological needs for self-professional image (pursuit of truth and service to the society). The internalization of professional regulations into self-discipline becomes the most important professional norms followed by journalists.

From the perspective of professionalism, the public expectations for the media are the requirements of professional standards where the media is objective, fair, accurate, reliable, correct and expert. These requirements are consistent with the empirical research findings of American scholars. The American public judges whether the media can be trusted from the perspective of professional standards. This book outlines the characteristics of professional orientation in the judgment dimension of the American media credibility.

Professionalism of Western journalism is superficial to the media itself, or is thought that in Western society "professionalism" is considered to be part of the bourgeois ideological hegemony which serves the establishment (Lu et al., 2002). It is difficult for the public to understand this operating mechanism. They superficially recognize that the media should be professional and judge whether it is trusted from the perspective of professional orientation.

In China the media has begun to be commercialized since the 1990s. It is generally believed that the media holds dual functions of "mouthpiece" and generating revenue. The "mouthpiece" is the most fundamental function of the media in China. For the ordinary people the media plays a role of government spokesperson, and their expectations for the media are more of the government functions.

With the "mouthpiece" function the public trust in the media is actually a trust in the government through the media. It is built on the trust of the political identity of the media. Media credibility depends on the government's credibility. Thus, the primary judgment dimension of media credibility in China is authority. For example, it was reported in recent years that farmers learned about tax reform and other policies from

Table 2.6 The most trusted newspapers among Beijing residents and the readership per week

Which newspaper do you trust most?			Cumulative size of readers per week (10,000 people)
Trust ranking	Newspapers	Selection ratio of the most trusted newspaper (%)	
1	Beijing Youth Daily	17.3	452.2
2	People's Daily	9.8	121.8
3	Beijing Daily	6.3	171.6
4	Beijing Times	4 6	307.2
5	Beijing Morning Post	4 3	265.1 1
6	Beijing Daily Messenger	2	155.3
7	Southern Weekly	1.4	59.7
8	Labor Morning Post	1	37.8
9	Beijing Modern Business	0.5	20.6

newspapers. They held newspapers in their hands to resist unreasonable and arbitrary charges by local administrative departments and believed that newspaper reports were actual government policies. This book concludes that the characteristics of judgment dimension of media credibility in China is a power-oriented, where power refers to the administration.

Under this perception of power orientation, ordinary people complain to the media when they are in trouble, and use the media as a feasible way to solve their problems. The media in China have established corresponding departments, such as the Department of Mass Work. This department receives inquiries from the ordinary people. In recent years the media with a high degree of marketization has a department or hotline that addresses the needs of residents. A reader's hotline of *Chengdu Business Daily* asked staff of relevant government departments to regularly answer hotlines and deal with complaints on the spot. For some things not related to the government departments, such as emotions, help, school information, etc. the media is the most popular channel for dissemination of news. As mentioned earlier, Liu Wenbao, a farmer from Hubei Province believes that the news media can solve a problem more quickly after reporting it in the local newspaper. There is a practical factor in people's trust in the media.

Because the dimension of media credibility in China is power-oriented, the administrative position of the media affects its credibility. The media at a high administrative level are supposed to have a correspondingly high authority and credibility. As a component of the vertical political power system, the media itself is distinguished from high and low administrative levels, or "near" and "far" from administrative institutions. Within newspaper groups the Party newspaper has the highest administrative level, and is "close" to political power. The metropolis newspaper is a sub-newspaper, and it must assume more revenue-generating functions. Correspondingly its administrative level is low.

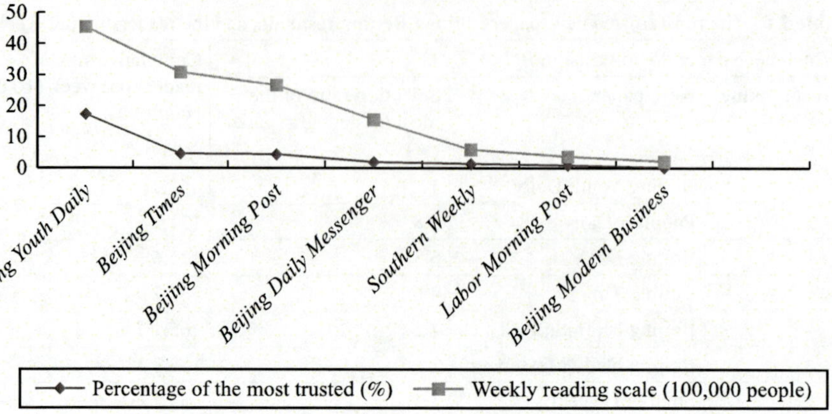

Fig. 2.2 Comparison of the newspapers trust ratio and readership

The Institute of Public Opinion of Renmin University of China conducted a survey among Beijing residents in September and October in 2002. We used the survey results and made further analysis of the data of "which newspaper do you trust most" and of "cumulative size of readers per week" from eight daily newspapers in Beijing (two are the Party newspapers and the rest are morning newspapers) and *Southern Weekly* which is influential among Chinese intellectuals. The data are shown in below Table 2.6 (Yu, 2003: 103–135).

We analyzed the newspapers other than two Party newspapers, i.e., *People's Daily* and *Beijing Daily*. Calculating the correlation coefficient between these two sets of data, we found that the Pearson Correlation Coefficient was 0.881. It was strongly correlated. In other words, the selection ratio of the most trusted newspapers among Beijing residents was proportional to their readership. The larger the size of readers, the higher the ratio of trust, as shown in Fig. 2.2.

But examining *People's Daily* and *Beijing Daily*, we found that their trust selection ratio and readership did not follow the above rules, as shown in Fig. 2.3. Their trust ratio ranked ahead of *Beijing Times*, *Beijing Morning Post* and *Beijing Daily Messenger*, which had larger number of readers. They were only behind *Beijing Youth Daily* with a much larger circulation. *People's Daily*, which is a Party newspaper, had a higher trust ratio than *Beijing Daily*, but their circulation was exactly the opposite. Finally, the number of readers of the Party newspapers, *People's Daily* and *Beijing Daily*, were smaller than those of their sub-papers, *Beijing Times* and *Beijing Morning Post* respectively, but their most trusted ratios were higher than the sub-papers.

Thus, media credibility is related to its administrative level. The fundamental point is that the dimension of media credibility is oriented toward administrative power because the authority of the power orientation is the most important dimension of credibility.

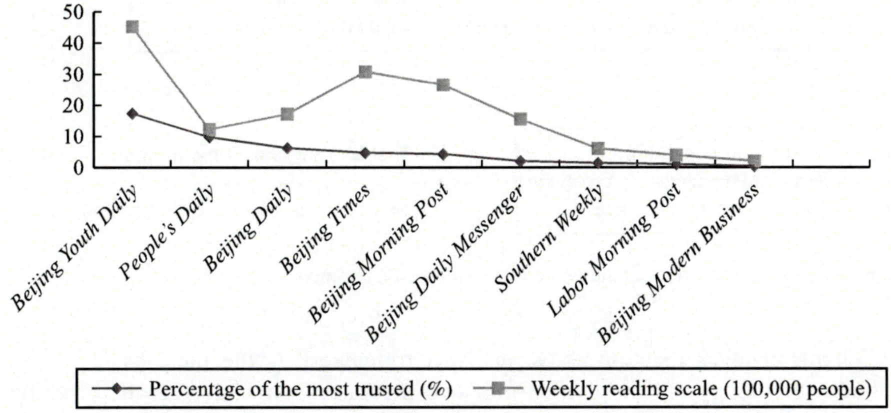

Fig. 2.3 Comparison of the newspapers trust ratio and readership with the Party newspapers

2.4 Value Analysis of Dimensions of Judgment

As discussed earlier, the judgment dimensions of media credibility in different regions are different, but different regions have the same judgment dimensions. Are the meanings behind these dimensions the same? That is the last question discussed in this chapter. Are there differences in the meaning of the same judgment dimension in different regions (or groups)? Are these terms such as objective and fair different in meaning to Chinese, Americans, Israelis, Palestinians, or people in other countries? It is concerned with the value orientation of judgment dimensions.

2.4.1 Two Dimensions of Judgment

In communication science some scholars use ethnomethodology to explore media credibility. Ethnomethodology was developed in the United States during the 1970s. It is a sociological approach to the study of practical actions that people take to accomplish their everyday lives. Professor Zhong (1992: 96) a scholar from Taiwan of China, explains that people have their own views on what is the nature of social phenomena and how society works as well as how society should work. So does the evaluation of media credibility. The public may not have a clearer idea of the media than academic discourse, but similarly they have their theoretical views.

Furthermore, from the perspective of cognitive psychology Professor Zhong (1992: 96–100) proposed a framework of ethnomethodology for analyzing media credibility. He argues that there are two dimensions in the judgment of media credibility, and each dimension (evaluation criteria) has its own basis. The framework is shown in Fig. 2.4.

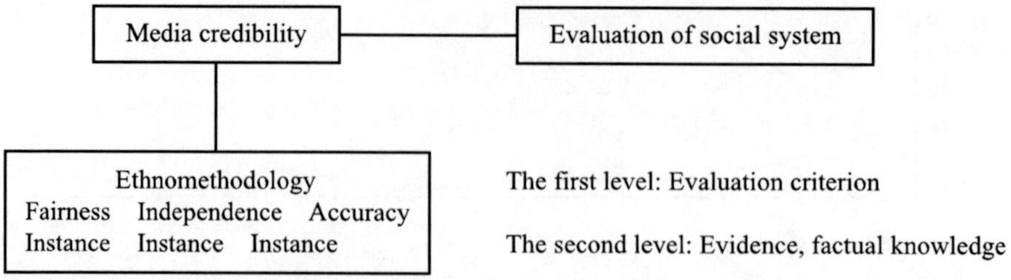

Fig. 2.4 The framework of ethnomethodology for analyzing media credibility

Zhong believes that the above analysis framework is the most basic one in ethnomethodology and is divided into two layers: the first layer is the criterion for judging the media; The second layer is evidence, which is like the process of constructing scientific theory. Individual evaluation of media is often based on some evidence and perception of media.

In this framework there are two psychological levels that people have to judge credibility. The first is the criterion for judging whether it is trusted or not, which is the dimension of credibility. This is measurable and on the surface. Under each dimension is hidden evidence supporting their respective dimensions. According to Zhong, whether a person trusts the media or not often depends on their views and their understanding of the media as a whole. The individual evaluation of the media is actually hidden and attached to the theory of ethnomethodology. Different people have different theoretical views.

This analysis framework provides a theoretical basis for deep analysis of the judgment of media credibility.

2.4.2 Dimension—A Kind of Value Analysis

In law and philosophy judgments are generally divided into fact judgments and value judgments. The fact judgment is the judgment of the actual condition of the object, and it is the judgment of "what" the object is. Value judgment is the judgment of the object on the meaning of the subject, and the judgment of "what the object should be". If the United States military overthrow the Saddam regime, this is an objective fact, and it is a fact judgment to believe that this matter is true. Some people say "yes", and others say "no". This is a value judgment.

From the perspective of ethnomethodology media credibility is a value judgment. Regarding the dimension of accuracy, different people may have different meanings to accuracy or inaccuracy of the same news report. In soccer games, for example, in a collision between players, the referee shows a yellow or red card to the players on both teams. At the press conference after the match, the coaches of both teams may say that the referee's penalty is incorrect, but their implications behind the same incorrect statement are different. They judge it from their own positions, and they think that the opponent player should be punished instead of their own player. There is a saying in our country that "both parties are not satisfied", which means that in the same way two parties get the same result that they don't recognize. The implications behind it are different. The same is true of whether the news report is correct or not. Even if people from different positions think it is correct or incorrect, the meaning of this judgment is different.

Regarding the dimension of biasedness, scholar Vallone et al. (1985) showed the movie news of the Sabra and Shatila massacre to two groups of people who supported Israel and Arab respectively, and measured their perception and attitudes towards the news. Members of two groups reached the same conclusion and thought that it was unfair, but their perception was quite different. They felt that the news was on the other party (Zhong, 1992: 99).

As for the dimension of justice, different individuals or groups mean different. For example, Zhong's research found that two respondents accused the media of being unfair, but their evidence was completely different. The first respondent accused the media of neglecting the bright side of society, while the second respondent accused the media of favoring the ruling party. Therefore, one couldn't conclude that they had the same attitude because the respondents chose the same answer and the same adjectives (for example, I believe the media is fair). This is an oversimplified reasoning (Zhong, 1992: 108).

Similarly, the other dimensions of the concept of credibility are the same as correctness and biasedness. The public have their own positions behind same dimensions.

It can be seen that the judgment dimension of the mass media credibility contains the meaning orientation of the subject, which means that media credibility is a value judgment. Putnam (1981: 128) argues: "The (fact-value) distinction is at the very least hopeless fuzzy because of factual statements themselves, and the practices of scientific inquiry upon which we rely on to decide what is and what is not a fact, presuppose values." From the perspective of the schema theory, a schema is subjective, and the same or similar media message generates different perceptions in the minds of different audience or individuals. The public judge the significance of each specific dimension of media credibility by their own schemas which are of the subjective value.

Since media credibility is a value judgment, is it irrelevant to judge the facts of the media? What is the position of fact judgment?

As mentioned earlier, the hierarchical schema has two characteristics. First, the schema at different levels have a ladder-like structure, and the schema at the next level are the basis for the schema at the last level. When the adjacent-level schema is missing, it will cause cognitive "faults" and cognitive block. Second, it is the combination of schema of various levels, which reflects the integration of the schema. When tens of millions of schemas formed in the recipient's mind are in an orderly state, which integrate, embrace and influence each other, constructing a huge cognitive network and system, the recipient is not bewildered and can choose what they want and digest them when facing complicated media messages (Liu, 1997: 137).

Thus, there is a hierarchical relationship between fact judgment and value judgment. Fact judgment lays a foundation for value judgment. The two are not separate but a whole. The public judgment of media credibility is about the media's description of specific events, and is the starting point of the judgment. Even if the value orientation of the media is recognized by the public, its value judgment cannot be realized if it doesn't follow the factual rules. Even if the credibility of the media is very high, the value judgment doesn't work if it clearly violates the principle of objective facts.

For example, in the "SARS" incident in Guangdong in early February of 2003, the media did not report on the outbreak at first. Until the morning of February 10th, the newspaper in Guangzhou issued a piece of news of four or five hundred words, which vaguely mentioned that the number of patients suffering from "cold" and "pneumonia" had recently increased in Guangzhou. The content of the manuscript is as follows:

> The reporter learned yesterday from major hospitals in Guangzhou that the number of patients suffering from cold and pneumonia in Guangzhou had increased recently. Experts told the public that viruses are prone to breed in spring. Citizens should avoid activities in densely populated areas. Prevention methods include burning vinegar at home to kill viruses and taking preventive antivirals.
>
> Zeng Jun, deputy director of the Department of Respiratory Medicine at Guangzhou First People's Hospital, told reporters that the temperature difference is big and humidity is high in Guangzhou in the spring. This kind of weather tends to weaken the human immune system, and especially human respiratory system is more vulnerable to be infected. The most common symptoms are cold and pneumonia. Respiratory diseases can be transmitted through the air, and coughing and sneezing can spread pathogens, so densely populated areas become high-risk places for spreading of disease.

The fact was that policemen on duty in public places such as Guangzhou Railway Station were wearing olive-green masks. It showed that the situation was much more serious. Therefore, although people have a high degree of trust in the media, when the media lacks some basic facts, the public do not trust it. As a result, the panic among residents in Guangzhou on this day reached a high point, and people rushed to buy the traditional Chinese medicine of Banlangen, white vinegar and masks.

After the basic fact judgment, the public will make a value judgment on the authority, accuracy, reliability, biasedness and integrity of media credibility, and then build the credibility of the media as a whole.

Chapter 3
Factors Influencing Mass Media Credibility

3.1 System Structure of Influencing Factors

The Functions of the Executive, written by the famous American management scientist Chester Barnard and published in 1938, inaugurated the school of social system in management theory. He held that the organization is a complex social system, so it should be analyzed and studied from the perspective of sociology and all kinds of organizations should be taken as cooperative social systems, as is the case with communication studies.

The literature review shows that the empirical school represented by American scholars focuses on the public and the media that are two factors affecting mass media credibility. Specifically, studies are conducted from the following aspects: gender, age, education, race of the public, the use of media, the dependence of media, differences of media traits, professional performance of media organizations, and other aspects. Their impacts on media credibility are discussed. According to the literature, there are some comparisons of phenomena on the longitudinal time axis and on the horizontal regional axis, but no special discussion has been found studying the impact of media credibility from the horizontal perspective of cross-cultural differences and cross-regional differences and from the vertical perspective of social systems and social changes. In other words, the current study mainly focuses on the analysis of variables on the micro-level, but variables on the macro-level are rarely considered, let alone the effect between micro variables and macro variables.

Therefore, this chapter is to analyze the factors affecting media credibility on each level systematically. As such, the media should be examined in a system.

The paper "Mass Communication and the Social System" was published in 1959 by John Riley and Matilda White Riley. It takes mass communication as a system among various social systems in society, and uses the sociological method to analyze mass communication.

The Rileys believe that as the communicator and receiver of the communication process, each individual is affected by the basic group. The communicator may be affected, and choose a special way to select and produce the message, while the

G. Yu, *Research on the Communication Effects and Mass Media Credibility in China*,
https://doi.org/10.1007/978-981-19-6242-4_3

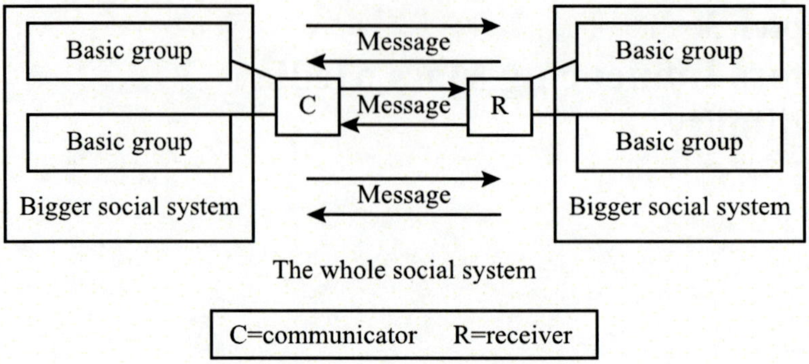

Fig. 3.1 The Rileys' model of communication system

receiver, facing the message, may be guided by basic groups when the one makes choices, comprehends and responds. Basic groups do not exist and act in the empty space in society. On the contrary, the Rileys see them as part of a bigger social system. A communication system is constituted in this way. However, this communication system exists in an all-inclusive social system. The process of mass communication affects the bigger social system and is also affected by it (see Fig. 3.1) (McQuail & Windahl, 1997: 47–50).

As for the sociological investigation of communication process, that is a representative model, which outlines the social environment in which the communication process exists and its relationship with the social environment. McQuail (1997: 47–50) comments that this model does help people link the concept of mass communication with existing sociological theories, and enables people to conduct the sociological analysis of research results that have not been adequately explained. It proposes that mass communication should be considered as a process among all kinds of social processes, and it affects the surrounding society and be affected as well. This view is of great significance.

The model of the Rileys made a systematic statement of the relationship between the communication process and the social environment, placing the entire communication process in the social system with different levels. The subject and object of the specific communication process, the communicator and the receiver, are in a certain social environment, and constitute a micro-communication system, including two sub-systems, a receiver system and a media system. The entire communication system is in a macro-level whole social system.

From a systematic perspective, the credibility of the media is affected by the macro social system, the micro receiver system, and the micro media system, respectively.

Hereupon, the influencing factors of mass media credibility are discussed from four aspects: (1) macro social system; (2) micro audience system; (3) micro mass media system; (4) influencing relationship between macro social system, micro audience system and micro media system.

3.2 The Influencing Factor of Macro Social System

3.2.1 Structural Analysis of Social System

This section is to analyze the structure of social system at the macro-level, and to find out structures that can be analyzed in the social system and the relationship between these structures on theoretical models, i.e. the effect of each structure in the social system, and then to analyze the influence of these structures on media credibility. Therefore, the social system at the macro level is analyzed first from the perspective of structural–functional theory.

Parsons is the representative of structural–functional theory. He first distinguished all aspects of all human activities with "systems". The total activity is an action system which includes three subsystems: the personality system, the social system and the cultural system. Personality system refers to all processes and characteristics of personal development, and it combines two elements of values and cognitive orientations. The social system is centered on the relationship between "characters in various roles" and the relationship between groups, and stipulates these relationships with a variety of criteria and social values. The cultural system focuses on expressing the meaning of things and defining moral connotation, and its components are beliefs, the concept of existence and symbols which show some concepts and moral values. Among three subsystems of personality system, social system and cultural system, Parsons believes that the cultural system plays a dominant role.

Parsons further analyzed the social system of human beings, and argued that it includes four parts: economy, polity, credit, and the societal community, among which credit and societal community constitute a cultural content in the social system and play a dominant role in it (See Fig. 3.2) (Walters, 2000: 162).

Economy

Investment	Production and marketing
Resource investment	Organization

Polity

Administration	Execution
Constitutionalism-judicature	Legislation

Logos	Goal
Secularism religion	Moral

Credit

Loyalty	Collectivity association
Execution	Solidarity

Social community

Fig. 3.2 Subsystems of social system. *Source* Parsons and Platt (1973); [AUS] Malcolm Waters edited

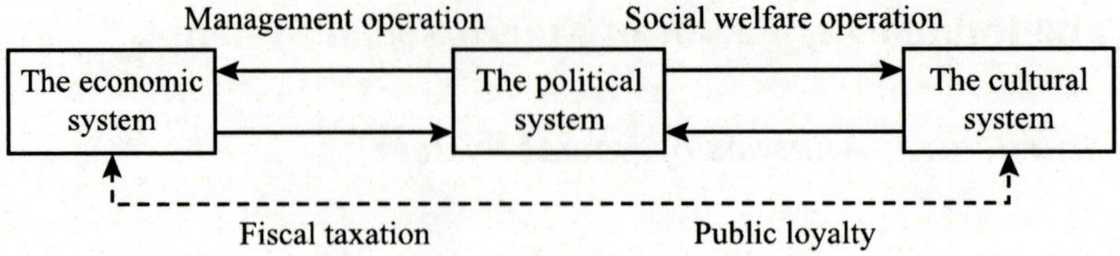

Fig. 3.3 Composition of the social system and its relationship

Here is a problem that in the action system of the first-level, the cultural system refers to the interpretation, classification, and institutionalization of morals, knowledge, and values, while in the subsystem of social system of the second-level, credit and social community refer to moral, religious, and the like. Is the relationship between the two progressive, or are they the one thing? According to the explanation of Parsons, it can be seen that the cultural system is the highest dominant level—the dominant system in the action system. It is abstract concerning culture, and it conceptualizes and institutionalizes the content of credit of the social system, whereas in the social system, the content of credit and social community is a kind of concrete and actual cultural content, reflecting the cultural system in the social system specifically. Since credit and social community belong to culture content, Parsons believes that it plays a dominant role in the social system.

For this problem in the social system, Habermas thinks that credit and social community in the social system should constitute a single socio-cultural system. As such, the social system should comprise three instead of four subsystems, which are the cultural system, the economic system and the political system. As for the relationship between these three subsystems, Habermas holds that the political system lies between the economic system and the socio-cultural system, but the latter two connect indirectly. The political system swaps management operation for taxation of the economic system, and public loyalty for the implementation of social welfare of the socio-cultural system (Walters, 2000: 176). Unlike the cultural dominance of Parsons, Habermas argues that politics plays the core role (see Fig. 3.3).

Therefore, there are three subsystems which are the political, the economic and the cultural subsystem in the social system. The cultural determinism of Parsons and the political determinism of Habermas are two theoretical models concerning the relationship among these three subsystems. For this study, the influence of mass media credibility on the macro-social level refers to that analytical model of social system, and it will be analyzed from three aspects, political, economic and cultural. Furthermore, the study aims to find out major factors influencing media credibility within these two theories about the macro-social system, and to discuss the effect and causes of major factors.

3.2.2 *Two Kinds of Determinism Affecting Credibility*

The above section analyzes theoretical models of the macro-social system structure, and two kinds of determinism proposed by these models, which are the cultural determinism of Parsons and the political determinism of Habermas. Relating to the research of credibility, which of macro-social variables, political or cultural, is playing the decisive effect? There are two kinds of determinism on this problem—the cultural determinism and the political determinism. Before the discussion about the influence of macro-social system on media credibility, two kinds of determinism on the influence of social system on credibility are first discussed in the following.

A prime representative of cultural determinism (Fukuyama, 2001: 27–49) is Fukuyama who argues that community depends on trust, and trust in turn is culturally determined. Spontaneous community emerges in differing degrees in different cultures. So what is culture? Culture is inherited ethical habit. Trust arises within a community of normative, honest, and cooperative behavior, based on commonly shared norms, on quality of members of that community. Those norms can be about deep "value" like the nature of justice, but they include real secular norms like professional standards and codes of conduct. That is, culture, shared by community, determines the occurrence of trust and its scale. Furthermore, from the perspective of Fukuyama, social capital is a capability that arises from the prevalence of trust in a society or in certain parts.

According to Fukuyama, culture in China, France and South Italy is low-trust, while in the United States, the United Kingdom and Japan is high-trust. With regard to the argument of Fukuyama, some scholars think that his conclusions are actually based on his personal impressions without any empirical basis. For this, Inglehart had conducted the World Values Surveys on dozens of countries for three rounds, and found that in general countries affected by Protestantism and Confucianism are more likely to generate trust than countries affected by Catholicism, Orthodox and Islam (Wang & Liu, 2003: 226). Although the conclusion is inconsistent with Fukuyama, it confirms the cultural determinism of Fukuyama.

Political determinism is held by some scholars of institutional school who argue that the existence of an effective government is crucial for the existence of trust. Hobbes perceives that a society without a public authority must be the one full of intrigues and massacre. Under that circumstance, people cannot trust each other. It is true that not all types of governments can help to promote social trust. The government should at least reward the reliable and punish the deceitful. It requires an integral legislative system under which the cost of dishonesty is greater, so people are more likely to trust others.

The theoretical question of Hobbes stems from the hypothesis that anarchy, absence of enforcement, leads to distrust and social disintegration. On the contrary, some scholars criticize that point of view and state that some empirical evidence shows the opposite direction. That is anarchy engenders trust. In other words, anarchy engenders social cohesion. It is effective government that destroys trust. The argument is that anarchy generates trust, but the government destroys trust (Gellner, 2003: 177–178).

Although both sides have their reasons to support their arguments in the above dispute, the key issue is that different governments need to be discussed differently. Naples in the eighteenth century is a common research object of trust among sociologists. After the Spanish armed forces went into Naples in the early eighteenth century, the monarchy was implemented, and trust there was destroyed. In order to reduce the wealth of Neapolitan citizens to a politically acceptable minimum and their political virtue to zero, the Castilian crown had set out to destroy the pre-existing bonds of trust within the society. This had been achieved partially by means which the French Kings had coped with their nobility. They had created large amounts of new nobles who were loyal to the king. In other words, it used a society based on honor to take place of one based on trust (Pagden, 2000: 127–141).

However, the republic helps to build trust. For Genovesi and Doria, trust (public trust) depends on the reliability of expectation. The only kind of society capable of generating and maintaining high levels of expectation which would render its people prosperous and happy was the virtuous republic. It is what Ciceronian called as *bene ordinate respublica* where all laws were directed towards the public good and the sovereign ruled in the interest of his people (Pagden, 2000: 127–141).

3.2.3 *Integrality and Dominance of Social System*

There are two meanings of the influence that the macro social system has on media credibility. First, the macro social system, composed by three parts, has its influence as a whole. Second, not every parts of the social system have the same influence. The following is the specific discussion about these two aspects.

3.2.3.1 Integrality of Macro Social System Influence

As analyzed in the above, three subsystems, political, cultural and economic, of the social system have their own influences on the media credibility.

First, politics, culture, and economy are components of the whole, and not isolated. According to materialism, the economic foundation is social relations of production during a certain period, and the base of the superstructure, such as politics and ideology. Politics is the intensive manifestation of economy, and its key issue is the state regime. Culture refers to the sum total of creative activities of human beings, mainly involving spiritual culture or its social ideology. The economic foundation determines its superstructure which in turn affects the economic foundation.

Second, mass media plays a role in three subsystems in reality. As pointed by Habermas when he discussed development stages of the public sphere, the newspaper entered into the political field from the cultural field, and then into the economic field.

Emerging newspapers that appeared in different places in Europe in the seventeenth and eighteenth centuries mainly transmitted information and criticized current affairs. During that period, the newspaper had been the guidance of public opinion, and newspaper publishers became communicators and conductors of public opinion. The business purpose of the newspaper played the second fiddle, and even did not aim to profit. However, with the rise of capitalism, private companies got involved into newspapers gradually, and came to control newspapers. Along with it, the news media went into the stage of commercial operation. Habermas argued that public relations and advertising affected news operation process, changing the business purpose of newspaper from the promotion of rational criticism at the early stage into the satisfaction of commercial interests. At the same time, the state power preformed an intervention in the public sphere of media. Therefore, the news media fell into being controlled by economic and political forces, leaving the single function of unifying ideology. As a result, equality and openness which were characters of the public sphere originally collapsed.

On the one hand, mass media such as newspapers, TV, and magazines are inheritors of social culture, and an important part of the social cultural system. Moreover, the media industry brought by the mass media operation is a large sector of the economic field. For example, in China, the media industry, surpassing the tobacco industry, has become the fourth largest industry. Mass media is the field for political contests all the time, no matter what kind of social systems the society employs, constituting the political system directly.

Therefore, political, economic, and cultural parts in the macro-social system inevitably have an impact on the media credibility. Nye (1998), Dean of the John F. Kennedy School of Government of Harvard University, used data to show that the credibility of mass media is affected by its external macro-social environment. He pointed out that evaluation rises and falls of the public on various organizations depend not so much on specific performance of the organization as on the political and economic soundness of the nation. For example, when the public trust in American journalism fell by half from 1965 to 1995, the public trust in universities, large corporations, and pharmaceutical industry also halved.

3.2.3.2 Dominance of Political Influence

Although the political, economic, and cultural sub-systems of the social system have their influences on the media credibility, degrees of those influences are not the same, which may be large and small, dominant and subordinate.

As mentioned above, at the level of the entire social system, there are the cultural determinism of Parsons and the political determinism of Habermas. Specific to the influence on credibility, there are also cultural determinism and political determinism. So, for the media credibility, which of the two subsystems, political and cultural one, plays the main role?

The previous discussion on judgment dimensions of media credibility touched upon impacts that the macro-variable of culture and of politics have on the judgment dimension. After comparing judgment dimension indicators of credibility in China and the United States, it is found that the judgment of credibility is governed by the political factor. That is to say, the political factor plays a decisive role in judging the media credibility. As such, the political factor is a dominant one in the macro-system that affects the credibility.

3.2.4 *Positive Correlation of the Political Factor and Media Credibility*

3.2.4.1 Arguments About the Influence of Political Factor

In the macro-social system the political factor plays a dominant role in influencing media credibility. However, there are different opinions about whether the dominant influence of political factors is positive or negative. Here is the controversy about the relationship between the decline of media credibility and the supervision of government over the media.

In 1999 Joseph Chan, a scholar in Hong Kong, China, discussed reasons for the decline of media credibility in Hong Kong, and he said that it was reported that the credibility of American media was declining, but according to the research conducted by Clement So, Chin-Chuan Lee and him, the credibility of Hong Kong media had also generally declined in the 1990s. Citizens as the audience attributed the decline in media credibility to the intensified media competition and the media's self-inspection. These were signals of danger. He argued that reasons of the decline could not be simply analyzed from the media control of the government and market competition, but should focus on the self-discipline of the media. Since without journalists, it is impossible to talk about observance and improvement of professional standards. Social culture activities should be promoted, and citizens' media capacity should be improved, so as to improve the media environment, and further to increase the credibility of the media.

A commentator in Hong Kong, China discussed this issue, and argued that another phenomenon worth pondering was that some people described the government as a flood or a beast and opposed the government-appointed organization to regulate the media. It seemed to assume that the media would die once regulated. However, Hong Kong Broadcasting Authority, the supervisor of Hong Kong electronic media, whose all members were officially appointed, had not let serious controversial accidents happen since establishment, and it was improbable to weaken the credibility of the media. According to the investigation of Chinese University of Hong Kong, the credibility of Radio Television Hong Kong which belongs to the civil service system, ranked the second place in the eyes of citizens and journalists, with 7.52 points and 7.55 points respectively, much higher than all best-selling newspapers. The commentator did not agree with the catastrophizing remark that the media would die once regulated either.

It can be seen that although the political system is a macro-dominating factor that affects credibility judgment dimensions, it is impossible to judge whether the influence on media credibility is strong or weak. Further comparisons are needed.

3.2.4.2 Positive Correlation of the Political Factor and Media Credibility

The relationship between the media and politics, based on closeness of the relationship, is summarized into three models: direct control, indirect control, and indirect influence of political power on the media. China, Singapore and the United States, taking as examples of three models respectively, are compared horizontally.

Direct control refers to the situation in China where the media is the mouthpiece of the Party and an extension of the government. The Party and the government directly control the media. That is, the media is a component of authority, and the government controls the media through the executive.

Indirect control refers to the situation in Singapore. The government controls the media through a combination of legislation and indirect administration. The system allows the government to exert considerable influence over the operation of mass media through legal control, political maneuvering and structural arrangements. For the sake of national security and public interest, the press is demanded to support the government in its efforts for national development rather than to challenge government policies… While the media environment in Singapore is highly restrictive, the Singapore government is not draconian in its enforcement of the restrictive press laws. Although the government did try to discipline the press by closing down newspapers and arresting offending journalists in the early years, guidance rather than direct control and censorship is more often applied today. The role of the press is defined as to help people of different races and religions create a sense of belonging by informing and educating Singaporeans of national policies and issues, and inculcating good values in the people (Hao, 1994).

Indirect influence mainly refers to capitalist countries represented by the United States. The government influences the media indirectly, while the media appears to be independent of political power. The media is parallel to political power, as a balance mechanism. Moreover, most of the media is run privately, and not subject to the control and influence of political power. In fact, the media cannot be isolated from politics fundamentally. Every private media organization has its own political interest groups, influenced or even manipulated by these groups. The most typical example is News Corporation. Murdoch supported the Conservative Mrs. Thatcher in three successive elections in the 1980s, and maintained a good personal relationship with her. His investment in politics received a generous financial return. After Mrs. Thatcher, he continued to support the Conservative Major, but when Major sought to get re-elected, Murdoch found that Blair of the Labor was more popular and more likely to succeed in the election. He changed his attitude to the Labor Party policy, rejected the pleading of his old friend Mrs. Thatcher, and resolutely decided to abandon Major and support Blair. The media that he controlled began to propagate Blair, and it was what Tony Blair urgently needed. They two catered to each other.

Now, let's take a look at the media credibility in three cases.

For Singapore's indirect control model and Western indirect influence model, Hao Xiaoming, a scholar at Nanyang Technological University in Singapore, made a discussion. The development press model pursued by the Singapore government is often challenged by Western communication scholars, who believe that development communication or journalism is nothing but a ploy of the leaders of developing countries to perpetuate their rule. Most criticisms of the Singapore press are based on the libertarian traditions of the West, which emphasizes the press's autonomy and freedom to criticize the government. Such criticisms are not surprising because Singapore's leaders have never intended to let the country's press become free, at least not reaching the Western standards (Hao, 1994).

According to Hao Xiaoming's investigation, to the question how much confidence the respondents have in local newspapers, 36.5% of the respondents said they had a lot of confidence, 53.6% of them had some confidence, and only 10% of the people said that they had very little or no confidence in local newspapers. Singapore was compared with the United States and the United Kingdom by the investigation, and Hao argues: "Despite Western criticism of the Singapore press, the local newspapers are doing pretty well in the eyes of Singaporeans. The overall results show that the majority of the respondents believed that the quality of local newspapers was good. At the same time, people expressed strong confidence in local newspapers. Compared with their counterparts in Britain, where only about 10% of the people trust the veracity of news reports, or the United States, where only about 14% of the people have strong confidence in those running the press, newspapers in Singapore are obviously more highly regarded by their readers." That is to say, media credibility of Singapore is higher than that of Britain and the United States.

Let's look at the comparison between China and the United States.

The following analysis is based on data from three surveys made by Chinese scholars in recent years, and through the comparison with the United States, the judgment of Chinese media credibility is made.

(1) According to *Research on Communicating Effect of the Media in Beijing's Bid for Olympics* in 2001 by Ke Huixin, two surveys on Beijing residents were made, involving media credibility. The topic of the survey was: Generally, your confidence on news reports. The results showed that in two surveys, "complete confidence" and "some confidence" totaled 85.3 and 91.2% respectively, of which "complete confidence" was 21.3 and 30.2%.
(2) *Survey on SARS issues among Beijing Residents* was conducted by Institute of Public Opinion of Renmin University of China. The survey was carried out from April 22 to 23, 2003, adopting the sampling method of RDD and visiting 396 Beijing households successfully through a computer-assisted telephone survey system. One thing to note is the investigation time. On April 20 of that year, the government announced the truth about SARS, so the survey was conducted

Table 3.1 Confidence of Shanghai adults (including 18 years old) in media

Unit: %

Confidence level	A lot trust	Some trust	Little trust	Not at all	Hard to say	Sum
	42.7	40.6	7.2	1.9	7.2	99.5

just two days later. Even though the media managed to conceal the epidemic situation previously, there was still 66.3% of people trust the mass media.

(3) In the later period of SARS, a research team of Tongji University made a survey from May 23 to 25 in 2003, on citizens aged 18 (including 18 years old) in Shanghai. The survey used DDA sampling method and a computer-assisted telephone query system, with a total of 431 valid samples. The result of the survey on media trust showed that "a lot trust" and "some trust" had 40% each, which totaled as much as 83.3%, while "little trust" and "not at all" sum less than 10% (Ding & Wang, 2003) (see Table 3.1).

Since survey items of China is different from those of America, indirect comparison can be made instead of direct comparison. The Roper surveyed American newspaper in 2001, and found that sums of "very trust" and "rather trust" were merely 36% in five scales, of which "very trust" was 13%. In three survey scales about trust in the press, "a lot trust" fell from 25 to 10% from 1973 to 2000, but "little trust" rose from 21 to 41% in the same period.

Another piece of data came from the basic social survey report of the National Opinion Research Center at the University of Chicago. Scholars Zhang Kewen and Hao Xiaoming, by sorting data of 20 years from 1973 to 1993 longitudinally, found that the credibility of newspaper and television declined significantly. In three scales, "much trust" of newspapers and that of television dropped from 23.4 and 18.8% in 1973 to 11 and 11.7% in 1993, and "little trust" rose from 14.9 and 22.1% in 1973 to 39.2 to 37.2% in 1993 (Zhang & Hao, 1995).

Based on the above data analysis, it is obvious that the majority of Chinese, up to about 90%, have basic trust in the media, of which a third trust it completely. However, on the whole, nearly 40% of American people hardly trust the newspaper, television, and the press. In other words, the level of media credibility among Chinese is higher than that among American.

Based on three places, it can be concluded that media credibility and political power have a positive relationship. That is, in countries or regions where the media is close to the political power, media credibility is high, while in countries or regions where the media is not close to the political power, media credibility is low.

3.2.4.3 Causes

According to Habermas, among political, economic, and cultural social system, the political one is core, and in terms of influence on the media, political influence is dominant. That is, relative to the economic and cultural system, the role of the political system played on the media is direct and intense. Different political systems have different requirements for the media, so the media plays different roles in front of the public. As a result, the public has various expectations for the media in different political systems. This point is analyzed in the previous chapter about two orientations of media credibility. Here is a simple cause analysis.

Because of different expectations for the media, people's expectations for media credibility are different. That is, the public has various judgment dimensions of media credibility in different political systems, as discussed previously.

In China, the media is the mouthpiece of the party, the public trust in the media is based on their trust in the media's political identity. The media, as a link in the chain of the political system under the current stable state in China, has relatively high trust.

In the United States, it seems that the media is independent of the government as a normal social component, so the public trust in the media comes from the expectation of its professional performance. However, the media is a special social sector in the society with high professional requirements from the public. It is impossible for the media to get rid of political and economic interference. In fact, the media is subject to political and economic manipulation which sometimes comes out inevitably and affects the public expectation for it. Therefore, the public trust in the media cannot be very high.

3.3 The Influencing Factor of Micro Audience System

3.3.1 Theories: Three Theoretical Models on the Audience

In short, the audiences are a collection of individuals who contact with the media. These individuals are widely distributed and numerous. Scholars in the field of communication have developed many theories on the audience about how to treat this collection, among which the theory of influence inconformity is widely used. By reviewing the literature about influencing factors of media credibility, it is found that the current discussion on influencing factors in the audience is basically based on three theoretical models of the influence inconformity theory, including individual difference theory, social classification theory and social relationship theory. The following is a discussion of three theoretical models (Defleur et al., 1995).

3.3.1.1 Individual Difference Theory

The theory of individual difference holds that individual differences of the audience cognitive structure is the key factor that affects their attention to the media and their behaviors on issues and things discussed in the media. With the same communication content and methods, there may have different communication effects on different individuals.

The theoretical basis is the theory of "stimulus–response" held by behavioristic psychology according to which human mentality and personality are mainly acquired by "learning", although they are affected by heredity. Through "learning", people form their own mental structures, including mental processes and personal mental characteristics. Mental processes refer to the cognitive process, emotional process and will process, which are often called knowledge, emotion, and will. Personal mental characteristics mean that individuals, in the process of socialization, are affected by the social environment, such as families, schools, parties and groups, and form their own personalities, such as unique interests, habits, and temperaments, and their psychological characteristics.

With different environments and experiences in the process of "learning", concepts and positions acquired by individuals are different, that is, individual mental structures are different. As such, differences in personalities and mental structures determine that attitudes and actions of the audiences are different. Specific to mass communication, neither uniform audience nor powerful media effects held by the "magic bullet theory" exist. In front of information provided by mass communication, each individual makes different choices and perceptions due to differences in mentality and personality, and consequent changes in attitude and behavior also vary from person to person.

3.3.1.2 Social Classification Theory

Social classification theory is known as social differentiation theory, social taxonomy theory and social category theory.

In the field of communication studies the theoretical significance of individual difference theory is that surpassing early theories, like the "magic bullet theory", which ignore the role of the audience and assume strong media effects, this theory began to pay attention to the initiative of the audience in the process of communication, and put forward audience selectivity and attentional comprehension. Its main limitation lies with the incomplete relationship between people presented and mass communication.

Whereas social classification theory believes that the audiences can form different social groups based on demographic similarities such as age, gender, race, education, religion and economic income. These groups, formed by the same and similar demographic factors, have similar personalities and mental structures, and share relatively consistent outlook of life and sense of value. Therefore, members of unified group have consistence in media selection, content exposure, and even responses to information. In this way, the audience can be divided into different groups for further research. Social classification theory is not limited to individual differences but emphasizes the unity within the group, and also pays attention to differences between the groups. This is why it is superior to individual difference theory. Individual difference theory focuses on individual personality and mental differences, while social classification theory on features of social groups. In a manner of speaking, the latter is the amendment and improvement of the former.

3.3.1.3 Social Relationship Theory

Scholars like Lazarsfeld found that some communication effects cannot be explained by individual difference theory and social classification theory. The communication scholar Elihu Katz pointed out that until recently, the social image in the mind of most communication scholars still consisted of atomized individuals, connected with mass media. The social "audience" was seen as a polymer of age, gender, and social class, and few people thought of the meaning of a rather informal relationship. The problem was not that communication scholars were not aware that the audience has families and friends, but that they did not believe these relationships affected campaign results. Therefore, informal interpersonal relationships were considered to be irrelevant to the modern social system.

Therefore, scholars such as Lazarsfeld and Katz believe that the social relationship of the audience has a huge impact on the audience. In contact with the media the audience is influenced by their informal social relationship as an important mediator when they make choices, interpret media content and take actions on it. That is, social relations can both strengthen and weaken the influence of the media. Social relations mainly cover interpersonal networks, group norms and opinion leaders, and specific to social relations of the audience, they mainly refer to their work units, social organizations and various informal groups. Social relationship theory provides a junction for mass communication and interpersonal communication, and the bridge linking two sides is social relations.

Another theory related to social relationship theory is group pressure theory which believes that group pressure can affect the audience's acceptance of media content. People tend to join groups that share similar opinions with themselves, and this opinion identification from groups strengthens individual assurance of these opinions. Once the information reported by the media does not conform to the group interests and norms, it can be resisted by the group. In this case, group members often doubt the media, and adhere to the original opinion. It is an indisputable fact that at this time the power of the media is weakened. If the conflict between media content and group norms is not particularly serious, the group can interpret the media opinion separately. Since it is closer to their original opinion, group members tend to accept this interpretation. At this time, the role of the media is weakened. Therefore, it is very difficult for the media to change people's established opinions, unless opinions from the media are in line with opinions of groups.

In general, three theories on the audience are not absolute, and each one may exist in the process of communication. As Defleur stated, each theoretical model acts on the audience's selection of the media to some extent. Intervention variables, representing different characteristics of the audience in these theoretical models, have a similar effect in the process of mass communication. Furthermore, Defleur analyzed the selective mechanism of these three models from the audience's attention, cognition, memory and action specifically.

3.3.2 *Variables of the Audience*

In the study of the audience influencing media credibility, scholars' discussions are mainly based on social classification theory and social relationship theory. Izard (1985), professor at the University of Ohio in the United States, believes that factors of the audience affecting media credibility are: (1) ideology of the audience; (2) party identity of the audience; (3) age of the audience; (4) geographical area; (5) race. In addition to factors listed by Izard, there are media use and media reliance. The following examines specific factors influencing media credibility, including demographic variables, media use, media reliance, media knowledge, social groups, party identity and political awareness.

3.3.2.1 Demographic Variables

Gender, age, race, education and other demographic variables that are based on social classification theory are popular variables in discussing the influencing factors of media credibility. Some scholars think that with regard to possible factors that affect credibility, most early American studies focused on demographic variables such as gender, age, race, education, residential area, personal income and family socioeconomic status of the audience. After comparing with their assessment of media credibility repeatedly, it was found that research findings are not the same. One of the possible reasons is that respondents of each study did not represent the same parent population, some of which took national adults as the sample range, some of which took residents of a certain area as interviewees, and some of which sampled school students (Lo et al., 1999; Wang et al., 1999).

Regarding the influence of various demographic variables on media credibility, researches conducted at different places and times do not have a well-agreed result. The following are research findings of specific variables.

(1) Gender. It is found by American scholars that men tend to trust newspapers while women trust TV (Carter, 1965: 29–34). However, Lo et al. (2003) from Taiwan of China finds that men tend to believe in TV election news, and thinks that men are more dependent on TV election news, which may be one of the reasons that men are more likely to believe in TV election news.

    In terms of the Internet, some American scholars investigated politically interested Internet users and found that the credibility of online messages had four factors including reliability, correctness, prejudice, and integrity, and gender was significantly related to them. In the mass, comparing to men, women trust the Internet much more (Johnson & Kay, 1998: 325–340). However, the scholar Ye (2000) from Taiwan of China found that men gave a higher evaluation for the credibility of the Internet.

(2) Age. For the credibility of the reported content, it was found by some Singaporean scholars that young and middle-aged people (30–49 years old) gave the lowest rating of whether the coverage of the ruling party and the government was fair and objective, young people (18–29 years old) gave a slightly higher rating, and respondents aged 50 and above gave the highest rating.

For TV, the American scholar Mulder finds that age and the credibility of TV are negatively correlated. That is, the younger the audience is, the more they trust. But for newspapers, results of various researches are different. Mulder finds that older people tend to trust newspapers, while scholars Lo Ven-hwei and Chen Shimin from Taiwan of China finds that on the contrary, young people would rather trust newspapers (Mulder, 1981: 635–638; Lo & Chen, 1993).

Johnson & Kaye (1998) conducted research on the credibility of online journalism from four aspects which are reliability, correctness, prejudice and completeness, and found that three of them were negatively related to age significantly. That is, the younger the audience is, the more they trust online news. Also, Ye (2000) finds that the younger the audience is, the more they trust online media.

(3) Education. Due to different samples and research methods, different researches have different results in terms of the relationship between education and media credibility. Lo et al., (2003) finds that people with higher education are less likely to trust election news on TV, and put forward that these people may be critical of content in TV, and less dependent on TV election news, so they give a lower rating for the credibility of TV election news. However, Mulder finds that the better educated think that TV news is rather credible, while the less educated trust newspapers.

(4) Income. The Singaporean scholar Hao Xiaoming finds that people with higher income tend to be more critical of the objectiveness and fairness of the coverage, and nearly half the people with a monthly family income of S$5,000 or more find the coverage not fair and objective.

Johnson and Kaye's research on the Internet users finds that income and online credibility have a negative relationship, that is, as income increases, respondents tend to judge online media less credible. But related research conducted in Taiwan of China finds that there is no significant relationship between income and media credibility.

(5) Other demographic variables. Izard (1985) finds that in America, respondents with geographic locations judge media credibility differently, that is, the closer the audience is to the political centre Washington, the lower they judge media credibility. Furthermore, racial differences have some effect on rating media credibility.

3.3.2.2 Media Use

Media use is a common factor in research on media credibility, and most scholars think that media use and media credibility have a positive correlation. According to Johnson and Kaye, for five online media such as online newspaper and online magazines, people would like to judge their preferred media as the most credible. Lo & Chen (1993) finds that in Taiwan of China, the use of newspaper is a significant variable for calculating newspaper credibility, and the use of TV news for TV news credibility.

Some research finds that frequency of use is not significantly correlated with media credibility. With secondary analysis of data from the American Society of Newspaper Editor's 1985 media credibility study Rimmer & Weaver (1987) finds that frequency of media use is not generally correlated with credibility, and frequency of reading newspapers is not significantly correlated with newspaper credibility. Watching a local or network television news is not significantly correlated with TV credibility; but time watching TV are significantly correlated with TV news credibility. News choice is significantly correlated with media credibility, and those preferring certain news on certain medium are more likely to rate the preferred news on the preferred medium highly credible.

3.3.2.3 Media Reliance

According to Defleur & Ball-Rokeach (1990), media reliance is that media system is assumed to be an important part of structure in modern society, and it relates to individuals, groups, associations and other social systems. The meaning, in part, of living in a society lies in achieving individual and collective goals, individuals, groups and large organizations must depend on other people, groups and sources controlled by systems, and vice versa. They argue that there are two cases of media reliance: daily reliance for ordinary information to satisfy basic needs, abnormal reliance for asking information and opinion on significant social changes.

Lo et al., (2003) studies the relationship between reliance on newspapers, TV, broadcast and the Internet of corresponds with media credibility, and finds that in terms of the credibility of election news, comparing to media use, media reliance is more predictable. Among the five media, media reliance is the most powerful variable for predicting the credibility of election news. It proposes two questions that need to be discussed further: the first one is that the relationship between media reliance, media credibility and media effect is worth studying by scholars who are interested in media reliance's impact on the audience; the second one is that this study finds that TV reliance is the most powerful variable for predicting the credibility of TV election news, but also has a remarkable predictive power for election news on other mediums such as broadcast, magazine and the Internet. The reason may be that TV is the most trusted and most dependent medium for the respondents. In contrast, fewer people trust and depend on broadcast, magazine, and the Internet, so even if respondents rely on election news reported by these three mediums, they may rely on TV to obtain election information. Therefore, it makes TV reliance predictive for the credibility of election news on broadcast, magazine, the Internet. The above argument is merely the author's inference, and the truth remains to be verified by future research.

3.3.2.4 Media Knowledge

Does the public's understanding of media affect their judge of media credibility? In this regard, the discussion made by scholars in Taiwan of China shows that the two are not correlated. Professor Lo Ven-hwei listed fourteen knowledge items related with the media in Taiwan of China. For example, "In which TV station is Li Siduan the news anchor?", "As far as you know, should the news in the newspaper be examined by the government before being printed?". For these items about media knowledge, the result of a regression analysis with credibility items shows that as for both newspaper and TV, media knowledge is not predictive for their credibility (Lo & Chen, 1993). So far, there has no research on media knowledge's influence on media credibility in China's mainland.

3.3.2.5 Social Groups, Party Identity, and Political Awareness

For the social relationship theory proposes that informal social relations affect the audience's choice of the media, the discussion of informal social relations in the research of media credibility is mainly centred on social groups and political parties. It is found that the audience's political parties and associations influence the judge of media credibility. As shown by Gunther (1992), judges of the same coverage differ significantly due to different groups of the audience. There are similar findings in Johnson & Kaye (1998: 325–340) about the manipulation of political stances. Scholars in Taiwan of China believe that these research results and the influence of party identity and political awareness on credibility can be explained by "the third-person effects": the audience tends to overestimate the influence of message which is contrary to their own position, and this perception will influence the credibility of message that the audience evaluates.

3.3.3 New Perspectives of Research Theory

The audience is an important factor for media credibility. Current empirical researches show that various demographic variables based on social classification theory have varying degrees of influence on media credibility, and even the same variable affect it differently in different regions. Media use, media reliance, party identity and social groups have influence on media credibility, but media knowledge based on the research of Taiwan of China has no relation with it.

From the above discussions, it can be seen that communication scholars have conducted in-depth research on influencing factors based on three theoretical models about the audience, and made a lot of empirical studies on these audience variables at different times and places. However, by analyzing the research literature, it can be concluded that the current discussion on the influencing factors of audience at a specific place and time is relatively fragmented, and is mainly based on Western theories of audience. Combining with the context of China, further studies can be discussed from the following theoretical perspectives.

(1) The theoretical foundations of the influencing factor of audience such as social classification theory and social relationship theory are derived from the study of western society. Ke Huixin and other scholars pointed out that communication studies came from the United States in the twentieth century when it was a mature industrial society. So far, main communication theories and audience research models have also come from Western society. Characteristics of Western media audiences are relatively similar, but are different from Chinese audience in social transformation period. Therefore, theories of mass communication, especially audience theories and research models, may be essentially incompatible with the reality of Chinese audience. To understand the current situation of Chinese audience, scholars should begin with the composition of Chinese audience.

    Ke Huixin and others analyzed differences between China and the West and pointed out that there was a fundamental difference between them. Considering stages of social development, types of western audiences are consistent, and they are often classified horizontally. That is, in the same social stage, the audience is classified by socio-economic status, gender, age, or lifestyle. For example, they can be divided into office workers, seniors, university teachers, and children based on their social groups, or into gender, age, education, occupation, income and other social research indicators.

    But in China, it is probable that audience differences exist in social development stages. At the current stage, the fundamental difference among the Chinese audience is reflected vertically. Although it seems that they are in the same era, in fact they are at different stages of social development. Therefore, they belong to different groups. Only based on the vertical classification is the horizontal classification meaningful. Ke & others (2002) further proposed that Chinese current audience can be vertically divided into three groups: traditional "group" audience, "mass" audience in industrial society, and "new group" audience in post-industrial society.

    Whether this classification of Chinese audience into three types proposed by Ke Huixin and others is reasonable or not needs to be verified specifically, but questions they proposed are very important. The theoretical foundation for analyzing influence of audience on media credibility in China is essential for its explanatory power.

(2) At present research on specific variables of influence still needs historical comparative analysis. By analyzing historical changes of specific variables,

scholars can find out their variation tendency and then some laws of social changes.

(3) Influences of these variables can be compared geographically. By comparing influences of the same variable in different regions, scholars may discover some laws under different macro-social contexts. In this way research on "points" can be linked into "lines", and by combining micro and macro perspectives, theories will become powerful in interpretation.

3.4 The Influencing Factor of Micro Mass Media System

According to the model of "mass communication and the social system" of the Rileys, the media system and audience system are sub-systems in the micro-level. There are three parts composing the media system, which are the media channels, media institutions and pressmen. Development of new technology in the media channels is separated as an independent part here. The performance of these parts of the media system affects media credibility directly, for example, whether the media abide by ethical standards, whether the media respect the public's right of privacy, whether opinions and facts are separated, whether certain unreported news that should be reported is hidden, whether the media is controlled by commercial forces, and whether the coverage is fair and just. The following is analyzed from four aspects of the media system: media channels, media institutions, pressmen, and development of new media technology.

3.4.1 Influence of Characteristics of Media Channels

For the same news content, the public has different perceptions if the content is reported through different communication channels. That is to say, characteristics of media channels affect media credibility.

There is much research on the comparison of TV and paper media. TV is direct and vivid, and it is easy for the audience to form a visual impression. As for the newspaper and magazine, audience can only receive information by reading them. Jacobson's research in the 1960s showed that TV credibility was significantly higher than newspapers. Multivariate analysis of TV and newspaper credibility shows that respondents believe that news on TV is more complete than that on newspaper. For this phenomenon, the researcher argues that the newspaper is a single-channel, visual-modality medium, whereas TV is two-channel, with auditory and visual modality. It could be that TV news, presented with sound, pictures in motion and even color, outclasses other media in simulating reality for reporting (Jacobson, 1969: 20–28).

This credibility difference caused by characteristics of media channel has been discussed by scholars. Newhagen & Nass (1989) argues that criterion for evaluating credibility of TV and credibility of newspapers are different. TV news is timely and vivid, and the judgment of TV credibility is mainly based on their specific report content, that is, it is based on the audience's perception of a series of scenes in news footage. However, the time and space between newspaper readers and editors are separated, so when readers judge the newspaper credibility, they tend to evaluate the institution, that is, people's judgment of credibility is based on the newspaper's role of an institution, not on their perception of a series of coverage. So far, the statement of Newhagen and Nass has remained a hypothesis, and no empirical research has been done to verify it.

The difference in public perception caused by media channel characteristics is a relative constant for media credibility. Its influence on credibility is relatively stable. In addition to channels of TV, paper media and broadcast, channel characteristics of the Internet are another type. Besides visual and auditory modality of TV, the biggest feature is interaction with the audience. On the Internet, interpersonal communication, small group communication and mass communication coexist. The different types such as BBS, websites, Web portals of traditional media and news websites have distinct influences.

3.4.2 Influence of Media Institutions

Media institutions are the decisive part in the value of the entire media system, as every manipulation or negligence in its operation may have a large impact on media credibility. The following examines its influence on media credibility from news rent-seeking and lagged management.

3.4.2.1 News Rent-Seeking

The term of news rent-seeking comes from the concept of power rent-seeking in economics and politics. It refers to media institutions which have the power of news reporting make the use of this power to seek improper political and economic benefits for themselves. There have many discussions about commercial rent-seeking.

Regarding the current mergers, acquisitions and expansions in the Western media, some scholars believe that news agencies are controlled by giant companies, where news, like other commodities, is only for sale, and this new combination has led to the collapse of journalistic standard of behaviour. One problem caused by the commercial invasion into the media is that the media independence is threatened, because controllers can use their power to achieve business goals, thereby reducing the media professional quality. As a result, the media lose its public trust.

A typical operation of news rent-seeking is that media institutions, purchased by commercial organizations, provide "paid news" for these commercial organizations, or for economic benefits, they coerce commercial organizations. It also happened to Chinese media in the process of being commercialised. On November 24, 2004, *China Newsweek* disclosed a roadmap of news extortion made by one Evening News.

Since 2003, out of creating income, one Evening News had reached an internal agreement that by exposing scandals it forced the parties to buy advertising, forming a "media corruption food-chain" composed of leaders of the newspaper, journalists, and coerced units.

Another operation, called "hush money", refers to that when the media receives adverse news on commercial institutions, they let the media block the information by providing financial compensation. Huang Wenfu, the chief editor of *China Business Times*, mentioned in a forum that the "hush money" has been a special income of some financial media, and he appealed to both companies and the media to take social responsibility. In short, when companies make negative news, they pay for the media to shut the media's mouth. It was reported that a well-known beverage company paid "hush money" of RMB600,000 in the form of advertising to a less-known industrial media in Beijing. "Hush money" is actually an open secret, a well-known "latent rule" in the media, not an exclusive phenomenon of financial media.

As for media institutions, "hush money" is a common way of using advertising for the right to speak. On October 24, 2003, *21st Century Talent News* published "Investigation on The Injustice Detention of Doctor Hu Kun". The report was about an online scandal one year ago that Doctor Hu Kun of Fudan University was criminally investigated because he was involved in a copy right dispute with his former boss Ping An Insurance Company. This incident was finally disclosed by the traditional media that acted as a more credible carrier. After this incident, it is worth paying attention to the conduct of the media, which is not a new topic.

It was reported by *21st Century Talent News* that a friend who contacted Hu Kun in September of 2003 told the journalist that Hu Kun was surprised that all of the media remained silent in the past two years, and none of them reported the incident publicly. But one man called a reporter of *21st Century Talent News* and revealed that it got much media attention, but in the end, even the most trusted newspapers in China was bought by Ping An Insurance Company through advertising fees. This kind of "transaction" and its credibility are not discussed here for the time being, but it cannot be denied that critical reports in exchange for advertising revenue has become a public secret in some media. Therefore, it is common that after publishing a negative report of some unit, a newspaper then carried an advertisement for the unit to "redeem the impact". Some media even suppress negative reports under the temptation of huge advertising fees.

These media institutions only focus on immediate interests, putting authenticity and objectivity of report aside, but the ultimate cost is the loss of media credibility in long term, or unable to build up its due credibility.

3.4.2.2 Lagged Management

Poor management of media institutions leads to coverage quality decline, as fake news, wrongly written characters, and wrong details are more likely to appear. As such, the audience would consider the report as untrue and inaccurate.

Fake news made by *The New York Times* is a typical example. After the exposure of Blair's fake news by *The New York Times* and Prague's deed, the newspaper established a special committee to examine its internal editorial policy. The committee was composed of 23 people, led by the deputy editor Siegel, and had three invited members who were Boccardi, president of Associated Press, Bird, the former inspector of *The Washington Post*, and Professor Wiggins of George Masson University. The first that the Special Committee inspected was the Blair Witch Hunt, the second was the signature of reporting, and the third was "head policy".

John Noble Wilford, the reporter of *The New York Times*, said that it was the poor communication between the superiors and subordinates that let Blair stay in the newspaper for a long time and be the reporter for domestic news. Besides, as discussed above, the problem was how to manage when new tools of communication come into use.

After the incident of fraud, related investigations show that only 46% of respondents believed that *The New York Times* was credible, which was much lower than the local newspapers (73%), Fox News (72%), and CNN (66%).

3.4.3 Influence of Pressmen

Besides characteristics of media channels and media institutions, the performance of pressmen affects media credibility directly. Pressmen include journalists, editors and related staff.

3.4.3.1 Difference Between Perceptions of Pressmen and the Audience

Although pressmen are able to recognize the importance of media credibility and maintain it in their work, their perception of media credibility is reflected in what kind of aspects is different from the audience's perception. The result is that the specific work of pressmen may endanger media credibility. *Examining our credibility: perspectives of the public and the press* was published by the American Newspaper Editors Association in 1999. In this report, six conclusions were drawn, which compared differences of how to understand credibility between the audience and pressmen. These differences can actually serve as the factor of communicator influencing media credibility.

For media credibility these differences between pressmen and the audience are an invisible damage to media credibility, because people involved cannot be aware of the harm brought by these differences, and it is easy for them to avoid investigations from various error correction mechanisms of the media institutions.

3.4.3.2 Satisfaction of Professional Requirement

The ABC of news reporting—accuracy, brevity, and clarity—is the basic requirement of journalism. The American scholar Melvin Mencher states that accuracy is reflected in information and language, brevity in writing the story, and clarity means there is no doubt about what happened. These are moral requirements as well as the necessities for the practice of journalism. Journalists take on the responsibility of informing people about the world around them so that they can act on what they read, see and hear. Their actions depend on clear and accurate information. In order to restore the status of the press as reliable information provider, reporters need to start with correct spelling of names, right address, real quotations and direct observation (Mencher, 2003: 70). That is to say, the basic premise for news reporting to gain public trust is that the news report itself must be accurate, brief and clear.

Reporters' understanding of specific reports deviates from the requirements of news itself. As stated by some articles, "news analysis and credibility are inversely proportional. When one goes up, the other goes down. Although it is widely recognized by the media industry that American journalists are becoming more serious and their educational levels are getting higher, but opinion polls show that an important reason for the credibility decline is that newspapers are increasingly inclined to analyse and interpret. This approach has changed the definition of news…".

3.4.3.3 Journalist Quality

The quality of a journalist refers to cultural taste and morality.

In terms of cultural taste, reporters should not pay too much attention to vulgar topics. The typical example is the paparazzo that specializes in digging the privacy of celebrities, and even their gossips. Scholars in Hong Kong, China pointed out that in recent years, some Hong Kong Newspapers had developed a so-called "paparazzo" culture, sending reporters to trail after entertainers and social celebrities and to expose their private lives. There is no doubt that it is the duty of reporters to discover news, but this kind of behaviour is questioned by some people about violating individual privacy and freedom extremely. When journalists enjoy their freedom of the press, they should also take into account other people's rights and feelings. After the rise of the "paparazzo" culture in Hong Kong media, endless conflicts between entertainers and reporters show that such extreme interview methods have caused a rebound, creating friction and mistrust between society and the media.

Morality of journalists includes reporting the truth, no fraud and no deception, respecting the readers, and no deliberate misrepresentation of the facts. As such, the Blair incident of *The New York Times* is a typical case of low morality. Similarly, Hong Kong media had an unfortunate incident of fake news. *Apple Daily* took the lead in luring a man to China's mainland to go whoring, and then wrote a story based on his experience. Later, some media followed suit, which was criticized by the society violently. As a result, Li Zhiying, the owner of *Apple Daily*, published a statement personally in the newspaper to show the willingness to apologize.

Some articles pointed out that the "paparazzo" phenomenon and frauds in Hong Kong media had greatly weakened the reputation and image of Hong Kong media among the public. A survey released by the University of Hong Kong in September 2003 stated that the public in Hong Kong SAR. rated media credibility with the score of 5.48 point which just passed. In July of the same year, the media received the score of 5.84 point, but the score in 1997 was 6.55 point, which showed that the public in Hong Kong SAR. were gradually losing their confidence in the media. What is even more shocking is that in the survey there were actually 38.8% of respondents thought the media was irresponsible in reporting; only 17.9% believed that the media was responsible, and most of them had no clear opinion.

3.4.4 Influence of New Technological Development

Development of new media technology has changed the original media ecology, which inevitably affects media credibility. The following is discussed from three aspects that may influence media credibility: the dilution of journalistic professionalism by new technological development, the challenge to traditional media management and multiple communication channels.

3.4.4.1 Dilution of Journalistic Professionalism Under New Technological Development

Requirements of traditional mass media like newspapers, TV, broadcast and magazines can be concluded into two levels which are basic skills and work standards, and the concept of journalism. According to statements of journalistic professionalism, journalism is an occupation, collecting, organizing, processing and spreading news. It is called a profession, because it refers to professional skills, work standards and evaluation criterion necessary for journalism which must be acquired through specialized training and shared by journalists. Therefore, people in the media industry often use the word "professional" and the word "amateur" to judge news, and give the former a positive meaning. The concept of "professionalism" goes far beyond the basic sociological characteristics of occupation. Additionally, it includes a set of beliefs about the social function of news media, a series of professional ethics that regulate journalism, a kind of mind that it is subordinate to the authority higher than the political and economic power, and an attitude to serve the public consciously (Lu & Pan, 2002).

With the development of Internet, the network has become an open platform where online forums, web pages and websites have changed the role of gatekeepers of traditional media such as newspapers and TV, as everyone is a communicator and can release news online. It is the use of blog that has taken this open way of communication into a new level. Also various communication tools, such as increasingly updated mobile phones, are able to send one message to many people at the same time, and are compatible with the Internet. To use communication tools based on these technologies, communicators do not need professional skills, and are not subject to professional standards. There are no punishment mechanisms for the release of general content. Since communicators are not professional, it is unlikely that they have the belief of professionalism. In other words, with the development of new technologies, journalistic professionalism has been challenged, and is going to be diluted.

In this case, good and bad communicating contents are spreading via these new technologies, and one of the consequences is the influence on the credibility. Moreover, the increase of various words of the news affects the public's judgment on the credibility of traditional media contents.

3.4.4.2 Challenge of New Technological Development for the Management of Mass Media

The application of network and communication technology in the editing of mass media such as newspapers, TV, and broadcast directly challenges the quality control of media editing departments.

Some scholar stated that within 48 h after the Lewinsky scandal on personal homepage, it appeared as the hard news in *The Washington Post* and *Los Angeles Times*. The scholar argued that in the following years, the influence (of the Internet on mainstream media) will hurt several professional abilities of journalists, such as their working ability, the ability to verify facts before publishing, the ability to pay attention to publishing necessity, and the ability to spread reliable and authentic information.

A direct example is the fraud made by Jason Blair of *The New York Times*. It was shown by the investigation of the incident that Jason Blair, a 27-year-old journalist, confused his readers and newspaper colleagues by a laptop and a mobile phone. Editors often received his e-mails, claiming that he was in the state of Maryland, Texas, or other distant places. When everyone thought that he worked hard outside the city, he stayed in New York and spun a yarn by himself. As he concealed his true whereabouts, he continuously entered into the newspaper's database and stole.

The emergence of new technologies such as networks and communication technologies has facilitated the fraud made by journalists and non-observance of professional operating standards. It is a challenge for mass media institutions to manage "speculators" effectively and to maintain news quality in the rapid development of new technologies.

Therefore, the journalistic professionalism is diluting on the communicating channels of new technologies; the widespread information release platforms and non-professionalism affect the public judging the authenticity of mass media content; and in the development of new technologies, the quality of media content management is challenged. They will directly influence the credibility of mass media.

3.4.4.3 Technological Development Accelerates Audience's Multiple Choices of Communication Channels

The development of new technologies such as networks and communication technologies influences the audience in the form of multiple choices of communication channels. The dependence on the original media reduced, and it is the reduced dependence that affects media credibility. The case of SARS in the spring of 2003 is taken as an example here to illustrate the diversification of media channels under the development of new technologies (Yu et al., 2003a).

First, the emergence and popularization of communication tools such as mobile phones and the popularity of the Internet have diversified information channels for people. According to a survey about how Beijing residents got information during SARS, there were many channels for people to obtain the information on SARS, including mass communication channels and communication tools which accounted for a large proportion, such as the Internet and mobile phones (see Fig. 3.4).

Second, in terms of age, the audience of new technology communication channels such as the Internet and communication is much younger, while the audience of traditional mass media is relatively older. The overall age of the Internet audience is the youngest, followed by phones, notice from work unit, newspaper, word of mouth, TV and broadcast (see Table 3.2).

Regarding the above phenomena, Yu Guoming et al. believe that from the perspective of development, the status of mass communication in the entire communication system will be challenged severely. One of the main reasons is the rapid development of emerging communication technologies and that various convenient and full-featured communication tools are constantly emerging, so people have many choices, and especially the emerging communication technologies is very attractive to the new

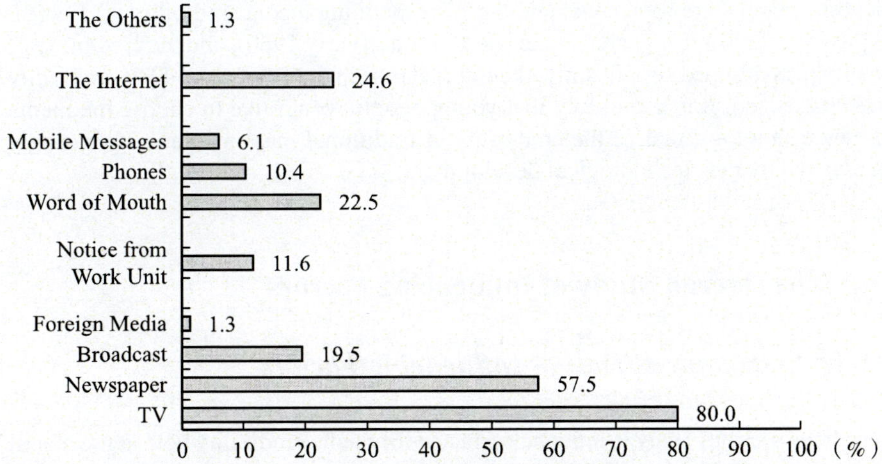

Fig. 3.4 Distribution of information channels for Beijing residents to know SARS

Table 3.2 Age structure of the audience of seven communicating channels Unit: %

Item	12–19 and below	20–29	30–39	40–49	50–59	60 and above	Total	Average age
Overall	6.5	26.1	22.7	18.0	12.8	14.1	100.0	39.31
The internet	8.2	44.3	28.9	11.3	5.2	2.1	100.0	31.31
Phones	—	37.5	37.5	15.0	7.5	2.5	100.0	34.50
Notice from work unit	10.9	15.2	26.1	23.9	17.4	6.5	100.0	38.73
Newspaper	6.4	25.1	24.7	18.3	15.1	10.5	100.0	38.81
Word of mouth	7.0	22.1	27.9	17.4	14.0	11.6	100.0	38.98
TV	6.5	22.3	23.9	19.7	12.0	15.5	100.0	40.02
Broadcast	3.9	23.7	15.8	17.1	17.1	22.4	100.0	43.24

generation. Statistics from this survey confirm that young and better educated white-collars tend to choose the Internet and phones; the audience of traditional mass media are relatively old; and the audience size of broadcast has already fallen behind that of the Internet. Second, the result of this survey shows that interpersonal communication has reached a considerable scale of which the scale of word of mouth surpasses broadcast. It is foreseeable that with the rapid development of Chinese economy, movement of population (at home and abroad) will definitely become more and more frequent, therefore the opportunity for interpersonal communication will increase rapidly and the spreading scope will expand with it.

The increase and diversification of communication channels inevitably affect the credibility of the existing media. The increase of communication channel adds a choice for the public, in which case the use and dependence of the original communication channel are weakened, and the corresponding media credibility is affected. Typically, television appeared in the late 1950s and early 1960s, and the credibility of newspapers dropped significantly; the Internet rose in the 1990s, and TV's credibility was challenged. It is a tendency that young people would like to choose the media of new technology, that is, the credibility of traditional mass media is increasingly challenged by new technological development.

3.5 The Overall Study of Influencing Factors

3.5.1 Systematic Effects of Influencing Factors

The above sections analyze influencing factors of media credibility both at the macro and the micro level, and the question here is that since these factors may affect media credibility, what kind of relationships do they have?

Based on the Rileys' model of "mass media and the social system", the audience and the media are in different levels of the social structure, influenced by different levels of the structure. Media credibility is in such an integral influence system, influenced holistically by every variable of the macro social system and the micro communication system, rather than a single variable or certain subsystem. The influence is systematic.

Judging from specific media, due to the differences of media orientation made by the media themselves and the differences of the audience, specific media influenced systematically may have one or some prominent influencing variables, so media credibility is of difference.

Historically, changes of the overall media credibility result from systematic influencing factors, rather than the effect of a single variable or part variables. This systematic influence echoes the changes of the overall society. The development and changes of the social system bring about a systematic change of influencing factors, which affects the overall media credibility. The credibility of the individual media is subject to the changing trend of the overall media. Although there may be individual differences, or it seems to be inconsistent with the general trend during a certain period, it is consistent with the historical general trend.

3.5.2 Downtrend of Media Credibility under a System of Influencing Factors

Historically, media credibility is affected by influencing factors systematically. So, what is the result of this systematic influence? Does it strengthen or weaken media credibility? The following is to make further discussion.

(1) *The historical change of the overall media credibility is studied.*

Since there is no continuous survey data of China, the United States is taken as the example to study the historical change of the overall media credibility.

The study uses the survey data of two institutions with the longest history of media credibility surveys, the Roper and National Opinion Research Centre of the University of Chicago. The data from the two institutions show that the credibility of the American mass media, regardless of the newspapers, magazines, or TV, or the trust in journalists, present a downward trend historically.

The Roper is a leading institution in investigating the credibility of mass media. This book sorted out the survey data on the trust of journalists from the Roper longitudinally, and found that from 1973 to 2000, the proportional number of "a great deal of confidence" in the press continued to decline. There are three options in the Roper's specific survey questions: a great deal of confidence, only some confidence, and hardly any confidence at all. "A great deal of confidence" dropped from 25% in 1973 to 10% in 2000, and "hardly any confidence at all" rose from 21% in 1973 to 41% in 2000. The change of "a great deal of confidence" is shown in Fig. 3.5.

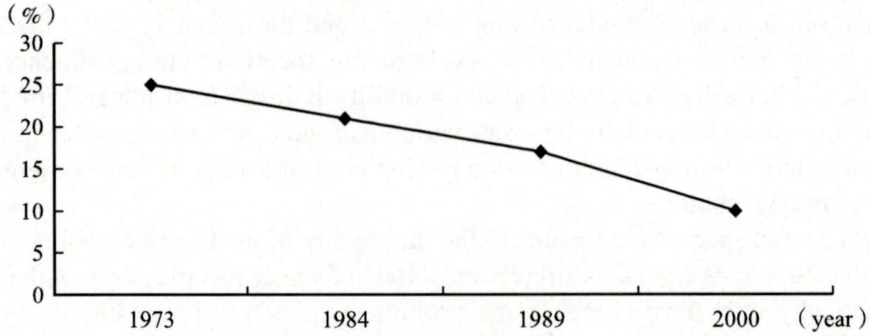

Fig. 3.5 Proportional change of "a great deal of confidence" in the press investigated by the Roper

Scholars Zhang Kewen and Hao Xiaoming longitudinally sorted out the data of the basic social survey report from 1973 to 1993 conducted by the National Opinion Research Centre of the University of Chicago, and found that the credibility of newspaper was higher than that of TV, rather than the crosscurrent argued by most scholars. This issue is not discussed here. The focus here is that both the credibility of newspapers and the credibility of TV had shown a significant declining trend in 20 years (see Table 3.3).

(2) *The systematic effect of influencing factors on media credibility is viewed from this phenomenon.*

It is obviously difficult for some audience, a variable of the media and the macro social variable to explain the overall historic downward trend of American media credibility. Most studies find that the educational level of the audience is inversely proportional to the trust of the media, that is, the higher the degree, the more distrustful of the media. However, this correlation depends on the specific media. If the TV has this correlation, it may not be obvious to the newspaper. Similarly, the influence of the technological development can explain the decline of some media credibility, and cultural diversity can too. The historical trend of media credibility is the result of the micro and macro factors.

In other words, specific micro variables or one macro variable may have great influence on the credibility of a certain media or a type of media at a particular time and place, so it is possible that some media have high credibility at some place in a certain period. But for the overall historic decline of media credibility, the influencing factor is not a single micro variable or the macro social system variable, but a systematic effect which is a combination of diversified systematic variables at different levels.

3.5.3 *The Systematic Negative Effect of Influencing Factors*

The systematic negative effect of influencing factors on American media credibility as mentioned above is affected by the macro social system, the audience system and the media system jointly. The following is a brief analysis of causes.

Table 3.3 Credibility of newspapers and credibility of TV (1973–1993)

Year	Media	A great deal of confidence (%)	Only some confidence (%)	Hardly any confidence at all (%)
1993	Newspaper	11.0	49.6	39.3
	TV	11.7	51.1	37.2
1991	Newspaper	16.7	54.9	28.4
	TV	14.5	55.0	30.4
1990	Newspaper	15.2	59.5	25.3
	TV	13.9	58.9	27.2
1989	Newspaper	17.1	55.5	27.4
	TV	14.3	56.3	29.5
1988	Newspaper	18.9	55.0	26.0
	TV	14.4	58.9	26.7
1987	Newspaper	19.3	57.4	23.2
	TV	12.7	59.7	27.5
1986	Newspaper	18.6	55.4	25.9
	TV	15.2	56.6	28.2
1984	Newspaper	17.3	59.9	22.8
	TV	13.4	57.8	28.8
1983	Newspaper	13.7	62.3	24.0
	TV	12.7	58.6	28.7
1982	Newspaper	18.3	61.6	20.2
	TV	15.5	59.3	25.2
1980	Newspaper	22.6	59.6	17.8
	TV	16.3	55.5	28.2
1978	Newspaper	20.5	59.5	20.1
	TV	14.0	54.4	31.6
1977	Newspaper	25.7	58.5	15.8
	TV	17.7	56.8	25.5
1976	Newspaper	29.0	53.0	18.0
	TV	19.1	53.2	27.7
1975	newspaper	24.5	57.1	18.4
	TV	18.3	58.8	22.9
1974	newspaper	26.2	56.1	17.7
	TV	23.7	58.8	17.5
1973	newspaper	23.4	61.7	14.9
	TV	18.8	59.1	22.1

Source Zhang and Hao (1995). Television Credibility Revisited: A Longitudinal Study. AEJMC Conference Papers, Washington, D. C. August 9–12

(1) At the level of macro social system, one characteristic of the changing American society is that social forces are constantly dividing and combining, becoming more and more diversified. The macro-social pluralism is manifested in politics, economy, and culture. For example, the end of racial segregation made ethnic minorities have certain political rights; the development of the new economy led to the emergence of new interest groups. As a Chinese legal scholar mentioned in the discussion of the US Simpson case, the case of Simpson is a wonderful lesson, so he learned what cannot be learned in the classroom. In this pluralistic society, interest groups and ethnic groups are becoming more divided, so it must be very difficult for the nation to reform consistently. In this changing environment, as a component of the social system, the influence of the mass media is affected by pluralism.

(2) At the level of individual, the education level of the public has generally improved, and values have become increasingly personalized. The personalized consumption is an example. B. Joseph Pine II and James H. Gilmore, authors of *The Experience Economy*, believe that with the emergence of large-scale customization, customers are increasingly focusing on personal experience and service feelings. Continuously accumulated wealth has provided a guarantee for a new life experience, which has nurtured an individualistic tendency more than ever before. The stronger, more complex and diversified feelings urge people to pursue a more personalized and autonomous life, and make the entire society pay more attention to the meaning of personal life and invest in various personal lives. This social change is a driving force turning the society of mass production into a new individualistic society which attempts to create a new world. The consequence of this personalization is that it is increasingly difficult for the media to meet the needs of the audience, which in turn affects its credibility.

(3) At the micro level, the media itself has problems. A series of fake news made by the mainstream media such as Blair's fraud case of *The New York Times*, mutual malicious criticism between the media, and the erosion by business have affected public trust in the media. Another big influence is the development of new technologies, as the diversification of media channels decreases media credibility inevitably. By examining data on media credibility of America, people can find several obvious nodes. As mentioned earlier, with the emergence and popularity of TV in the early 1960s, the credibility of newspapers was significantly lower than that of TV. After the mid-1990s, the credibility of TV and newspapers was affected by the widespread use of the Internet. When new technology emerges, the public re-examine media credibility. At present, besides network technology, the development of communication technology is changing rapidly, and mobile phones and telephones are very common, so there are more and more communication channels that people can rely on. Advanced and convenient social transportation makes interpersonal communication frequent. Due to the rapid development of network and communication technology, the public has diversified communication channels and can receive messages easily.

This situation of American society is a reference for other countries and regions.

The systematic influence of media credibility comes with the development of society. Although the development of the entire world is uneven and there are obvious differences between developed and backward countries and regions, every country is in its rapid development trend, faster than ever before. The result of development is the improvement of education, the increase of communication tools, the enlargement of interpersonal communication, and the three-dimensional social structure. All of them help individuals judge independently. When citizens of a country become knowledgeable and more sophisticated, their thinking styles become increasingly independent and their minds become more and more "obstinate". Everyone has opinions, so it is hard for the people to be persuaded by a certain point of view or a certain position, let alone a certain editorial (Du, 2004). This statement may not be theoretical, but it proposes that the ability of individuals to think and criticize independently with the overall social development affect media credibility. This point is worthy of our attention. In face of the overall social development, media credibility of every country and region is affected more or less.

In theory, the systematic influence brought by social development on media credibility will continue to play a role in the coming period.

In terms of the variable of age specifically, one article in *Washington Observer* stated that experts were concerned that the news consumption by the young in the United States hit a record low. Of the respondents under the age of 30, only 23% of them said that they often read newspapers; 18% watched evening news, 29% watched cable news, and 36% got information from the Internet.

The analysis of most survey data shows that the younger the age, the more frequent the use of new media and the lower the use of traditional newspapers, TV, broadcast and other mass media; and most of the data also support that the younger the age, the less trust in mass media.

Therefore, judging from the general trend and specific variables, with the development of society, it is foreseeable that the credibility of mass media may be affected by systematic factors more than ever before. Several issues are worthy of discussing in the future, such as the degree of such systematic influence on mass media, and how to deal with the influence.

Chapter 4
Generative Mechanism and Control Analysis of Mass Media Credibility

4.1 Generative Pattern of Media Credibility

To explore the generative mechanism of the credibility of mass media, we will first examine the generative mechanism of trust, and then establish the generative pattern of media credibility in this research.

Previous review of research shows that scholars have different research models on the mechanism of trust generation, which are proposed from their disciplines and research perspectives. After a comprehensive analysis the scholars' study in different disciplines, this book divides the generative mechanism of trust into two dimensions, i.e., horizontal generation and vertical development. One is to examine the model of trust generation from a horizontal perspective, and another is to discuss the model of trust development from a vertical perspective.

4.1.1 Horizontal Generative Pattern of Trust

The typical ones are the production model proposed from the sociological perspective and the game theory model by economists. The two models are discussed below.

4.1.1.1 Sociological Model of Trust

As for the question of how to gain trust, the representative is the generative mechanism of trust analyzed by Zucker from the perspective of sociology in the Western trust research. Zucker identifies three types of trust: process-based trust, characteristic-based trust and institutional-based trust (Peng, 1999).

The process-bases trust is developed through repeated interaction and exchange of interests, such as exchanging gifts. Reciprocity is at the core of this process. In addition, participation in the process can generate reputation-based expectations,

G. Yu, *Research on the Communication Effects and Mass Media Credibility in China*,
https://doi.org/10.1007/978-981-19-6242-4_4

and decide whether to give trust based on knowledge of others' past behavior and reputation. In short, the reliability and stability of repeated reciprocal exchanges enables participants to learn while fostering interpersonal trust. In Zucker's view, this process-based trust refers to a reputation-based trust, that is, the trust is established by the credibility in the process.

The characteristic-based trust is based on the cooperation cultivated by voluntary norms and social similarities. When deciding whether someone is reliable or unreliable, we often take into account his or her family background, age, social status, economic status, ethnicity, etc. In general, with more similarities the trust is higher, for similar social backgrounds often mean similar behavioral norms, and it is for us easy to understand each other, and to reach consensus in exchanges.

The institutional-based trust is closely linked to the formal social structure. It is judged whether an individual has the attributes of the institution, such as trust given by professional qualifications, bureaucratic organizations, intermediaries and various regulations (Greed & Miles, 2003: 24).

Zucker used this model to analyze trust in American economic activities from 1840 to 1920, and found that with the influx of large numbers of immigrants and the intensification of population movements, the effectiveness of process-based trust was reduced. However, due to the development of rationalized bureaucratic organizations, the promotion of professional qualifications, the strengthening of regulations and legislation, the mechanism of trust generated by the legal system has been widely used, and the institutional-based trust become more widespread. Other scholars use Zucker's analysis model to examine the Chinese social trust generation mechanism. Limlingan, Whitley, and Yoshihara argue that trust in Chinese society is mainly generated by reputation and relationships, while legalized trust is rare.

Chinese scholars believe that the three types identified by Zucker fail to explain the main mechanism of trust in China. In their opinion, in addition to Zucker's three types, there is a mechanism for establishing interpersonal trust and relationship, that is, the activity of establishing, developing, maintaining and utilizing relationships. Considering the unique importance of relationship in Chinese society, Chinese scholars believe that relationships may be the main mechanism for Chinese people to build trust.

Professor Yang Zhongfang thinks that in social communication, the main function of relationship is that it guarantees the trust required at all stages of communication. Relationships imply mutual obligations, and a sense of obligation makes people act in a trustworthy manner.

Reciprocal obligations are a core element of the relationship. If a person fails to fulfil his obligations, he or she will lose face, who will be blamed by others and has to pay a great price—losing the network and the social resources in it. Since obligations implied in the relationship have such a large restrictive effect on individual behaviors, the trust building can be achieved through the establishment and development of the relationship (Peng, 1999).

Peng (1999) believes that the generative mechanism of trust varies from culture to culture and changes over time. The importance of relationship in trust building among Chinese people reflects the influence of culture. The change in the mechanism of trust generation in the process of American industrialization displays the influence of changing times. At present, China is in a period of social transformation, and this transformation has a certain impact on the trust-building mechanism of the Chinese people. Some major social characteristics of the American industrial phase mentioned by Zucker (such as the intensification of population movements, the strengthening of regulations, etc.) have appeared in the current Chinese society, thereby legalized trust may be strengthened correspondingly.

In summary, there are four types of trust generation: process-based trust, characteristic-based trust, institutional-based trust and relationship-based trust.

4.1.1.2 The Game Model of Trust in Economics

Economists have constructed game models to explain the generative mechanism of trust (Zhang, 2003: 35–38).

The game mechanism supposes two parties, one is the principal and the other is the agent. Clients can be called as customers, audiences, employers, etc. Agents refer to businesses, media, managers, etc. Thus, in the first stage of the game, if the principal does not trust the agent, the transaction cannot proceed and the income of both parties is zero. If the principal chooses to trust, the game enters the second stage, and the agent is required to choose an honest or deceptive one. If the agent chooses to be honest, each of the two parties will receive half of the payoff respectively. If the agent chooses to deceive, the agent receives zero income, and the client loses half of payoff (see Fig. 4.1). In Fig. 4.1, the first number indicates the income of the principal, and the second indicates the income of the agent.

It can be seen from Fig. 4.1 that for one-off games, agent deception can get the most benefits. However, in order to gain long-term benefits, it is necessary to establish a repeated-game model so that the game can continue. This payoff relationship can be shown in Fig. 4.2.

As can be seen from Fig. 4.2, it is clear that this kind of game where the agent chooses to cheat cannot be repeated. Deception can only get the maximum payoff (A) once, but cannot receive the larger and longer-term payoff (B). In this way the parties are willing to resist the temptation of one-time immediate benefits brought by

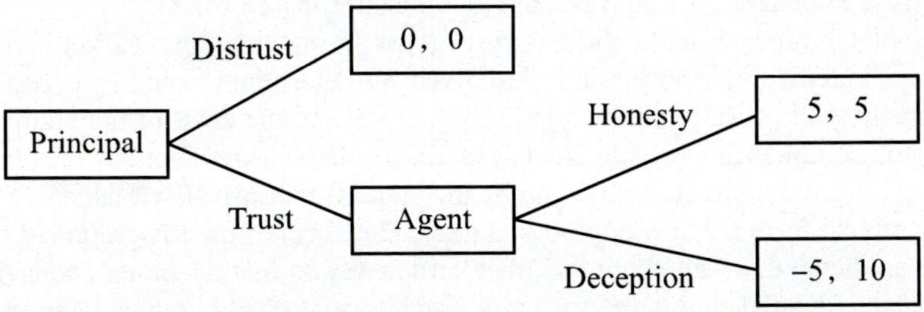

Fig. 4.1 Trust game. *Source* Quoted from Zhang (2003)

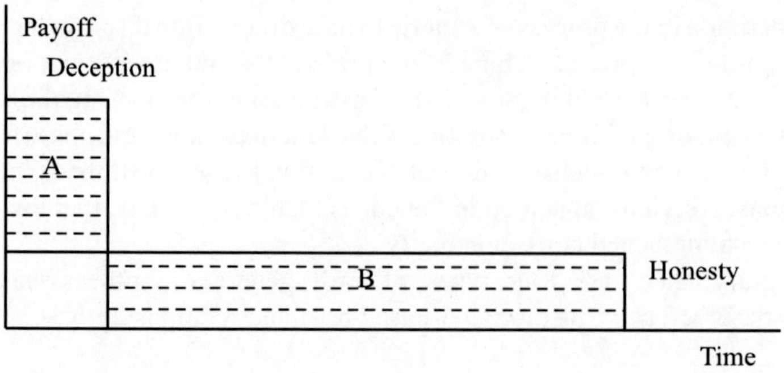

Fig. 4.2 Current payoff and future payoff. *Source* Baker et al. (2000); Quoted from Zhang (2003)

deception in order to benefit from the long-term benefits of cooperation and choose an honest game mechanism. In a market economy, companies turn one-time games into the mechanism of repeated games, which is the carrier of credibility. In this repeated game, trust is established.

Unlike social scholars such as Fukuyama, trust is more dependent on culture. Economists believe that although it is related to culture, good trust often results from the people's rational choice. In the repeated-game model, economists conclude that the pursuit of long-term payoff leads to trust.

4.1.2 Vertical Development Model of Trust

The above introduced the trust generation model in sociology and economics from the horizontal dimension. The following examines the model of trust development from the vertical dimension. The model of trust development is typically a three-stage model of trust development proposed by Lewicki and Bunker in 1996 (Lewicki & Bunker, 2003: 151–163).

Based on Shapiro et al. (1992), Lewicki and Bunker proposed a model of trust development, which suggests that trust has at least three components that operate in a predictive and developing way in the professional relationship. They are calculus-based trust, knowledge-based trust, and identification-based trust.

The first kind is calculus-based trust. Trust is an ongoing, market-oriented, economic calculation whose value is derived by determining outcomes resulting from creating and sustaining the relationship relative to the costs of maintaining or serving it. Compliance with calculus-based trust is often ensured both by the rewards of being trusted (and trustworthy) and by the "threat" that trust is violated.

The second form is knowledge-based trust. This form of trust is grounded in the other's predictability—knowing the other sufficiently so that the other's behavior is anticipatable. Knowledge-based trust relies on information rather than deterrence. It develops over time, largely as a function of the parties having a history of interaction

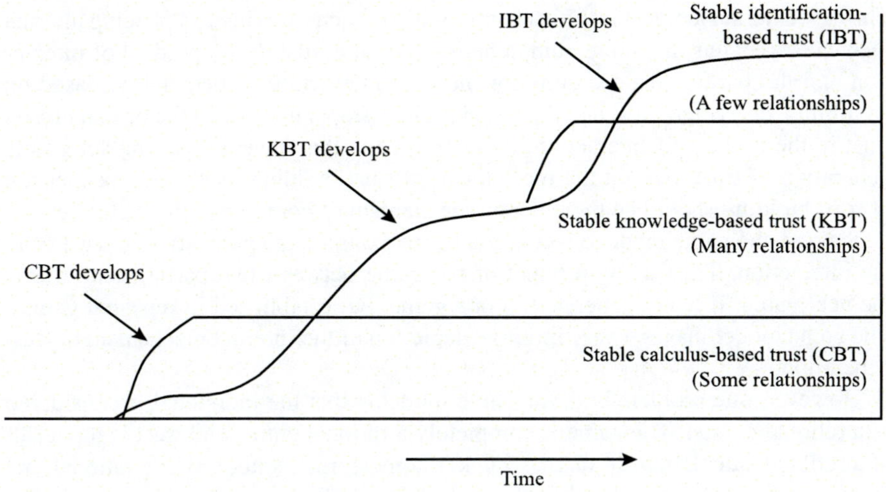

Fig. 4.3 The three stages of trust development

that allows them to develop a generalized expectancy that the other's behavior is predictable and that he or she acts trustworthily.

The third type of trust is based on identification with the other's desires and intentions. At this level, trust exists because the parties effectively understand and appreciate the other's wants; this mutual understanding is developed to the point that both parties can effectively act for the other. The three stages of trust development are shown in Fig. 4.3.

4.1.3 Generative Pattern of Media Credibility

4.1.3.1 Models of Trust Generation

First, trust is taken as the starting point of social activities (that is, trust precedes social activities). Zucker proposed three models of trust generation from the perspective sociology. Sociologists in China have summarized the major forms of trust from the perspective of relationships. That is to say, trust is obtained from four types: reputation-based trust, characteristic-based trust, institutional-based trust and relationship-based trust.

Which comes first, trust or exchange? Is there trust from the beginning? Is trust based on credibility, characteristics, institutions and relationships, or based on interaction? Obviously, trust arises from communication and has impact on it. The four models of Zucker et al. go beyond the interaction to discuss the mechanism of trust generation. The second issue is that credibility, characteristics, institutions, and relationships are the same variables as age, gender, education, income, and occupation.

They are external factors that affect communicative activities instead of being internal mechanisms. Thus, it's more comprehensive to take trust as the product of interaction and the parties' expectations for the next interaction. Whether it is based on credibility, characteristics, institutions or relationships, it is only a factor that affects trust in the process of interaction, or some factors that have a great impact on the generation of trust, but not the mechanism of trust building itself. The mechanism of trust building should be the inherent mechanism of communication activities.

Second, the game model of economics is to examine the generation of trust from the interaction. Trust is the product of the game between two parties, and affects the next game. It is not generated in one game, but established in repeated games. The game model makes a very intuitive depiction of the internal mechanism of trust generation.

However, one problem with the game model is that the individual is considered as a rational person. The game is completely a rational game, and trust is the result of a rational calculation. In fact, as Zucker argues, trust is not a single rational, but there is emotion, belief and other elements in it. In the process of the game of trust, these elements are mixed. In other words, the single rationality of the game model cannot fully explain the psychological contents of trust.

In addition, the vertical development model of Lewicki and Bunker divides the level of trust well. Trust is a deep and shallow concept in vertical development. From rational calculation to emotional identification, this model makes a good explanation to the vertical development of trust. The vertical trust development model of Lewicki and Bunker well addresses the single rationality of the game model. Combining the two models can better explain the generation and development of trust.

4.1.3.2 The Analysis Model of Media Credibility Generation—Frame Game on a Multi-dimensional Level

First, trust is the logical starting point of the concept of credibility, and the basic model of interpersonal trust generation is applicable to media credibility. They differ only in some specific contents. If trust is generated through repeated games, this mechanism is the same for credibility and interpersonal trust. They differ when the subject and object are different. The subject and object of interpersonal trust are two individuals, while the subject and object of media credibility are the public and mass media. Another difference is that benefits from the game of interpersonal trust may be long-term cooperation, money or power, etc. The benefits from the game of media credibility may be reliable information, entertainment, etc. for the public. The benefits from the game of media is to improve audience ratings, readership, etc., and finally to create influence and good returns of commercials.

Based on the previous research on the generative pattern of interpersonal trust, the analysis of the generative mechanism of mass media credibility makes use of the three-stage model of trust development proposed by Levitsky and Bunker in the overall vertical development: calculus-based trust, knowledge-based trust, and identification-based trust. As for media credibility, the corresponding stages are calculus-based media credibility, knowledge-based media credibility and identification-based media credibility. For each development mechanism in the horizontal direction, a game model is used for analysis. Especially in the first stage of calculus-based trust, a typical rational game model is used. With the combination of horizontal and vertical models, a single-dimensional rational game model is placed on different levels of development dimensions, from rational calculation to knowledge, and finally to emotional identification.

Second, for interpersonal trust, the subject and the object are both individuals, and the trust generation mechanism between them is analyzed, with the individual as the unit of analysis. But for the credibility of mass media, the subject is the public and the object is mass media. The public and mass media are not as simple as individuals. The public is a collection of many individuals. Mass media is a complex of individuals (reporters, editors, etc.) and organizations. Its composition is more complicated. So, how do the public and mass media play games? More directly, in what way do mass media and its audience play games?

The perspective of frame theory is used in this book to analyze the subject and object of media credibility, that is, the media and the audience participate in the game with the media frame and the audience frame respectively. The following is an analysis of the development stage of credibility by examining the game between the audience frame and the media frame.

For the generative mechanism of mass media credibility there are three stages in the vertical development: calculus-based media credibility, knowledge-based media credibility and identification-based media credibility, or simply, calculus stage, knowledge stage and identification stage. The mechanisms between the subject and the object at each stage are game models. The subject and object are not a single individual, but a relatively complex. The game between the audience frame and mass media frame. In this way the book proposes the generative pattern of media credibility as shown in Fig. 4.4.

4.2 Generative Mechanism of Media Credibility

The generation of media credibility is a repeated game between the media frame and the audience frame. First, the meanings of the media frame and the audience frame as well as their relations are discussed. Second, how the credibility is generated in the game will be examined.

4.2.1 Media Frame and Audience Frame

The concept of frame originated from Bateson, and was introduced into cultural sociology by Goffman who thinks that what is real for a person emerges in the person's definition of the situation. This definition can be divided into strip and

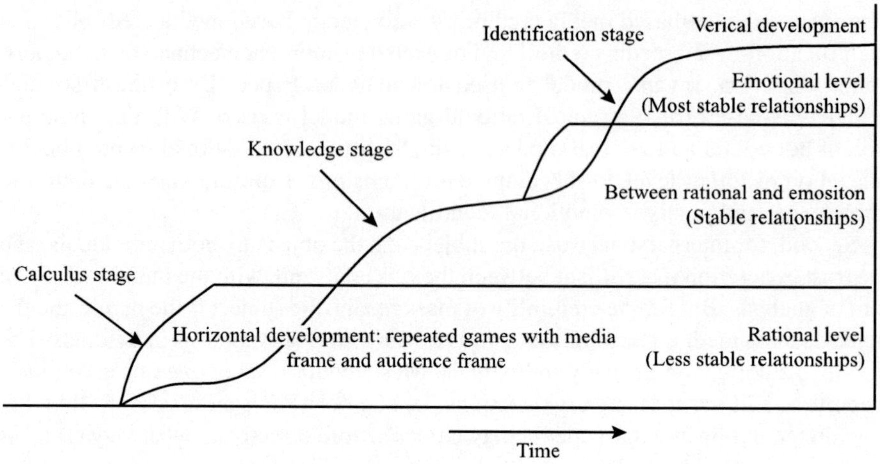

Fig. 4.4 The generative pattern of media credibility

frame. A strip is any arbitrary sequence of activity. A frame is a basic element of organization used in defining a situation (Littlejohn, 1999: 300). Simply put, the frame defined by Goffman is that people transform the reality of society into schemas of subjective ideas. According to Zang (1999: 26), a frame is the subjective explanation and thinking structure of people or organizations for an event. It is a set of principles for understanding social behavior or handling things. People use the frame to construct the meaning of a situation. So the frame is the setting, perception and identification of events or information by individuals or institutions. So how does the frame come about? Goffman argues that the frame is derived from past experience, and is often subject to social and cultural values.

The research on media frame and audience frame is based on Guffman's thought. The following discusses the media (news) frame and audience frame.

Zang Guoren is one of the Chinese scholars who has conducted in-depth theoretical discussions on the media frame. He believes that the media frame is the basic model of thinking that news media or journalists rely on when processing meaningful information. It is the basic structure of explaining external things. The media frame is actually an activity of constructing meaning. Journalists convert the original event into a social event, and after considering the public nature and social significance of the event, they convert it into a news report. In this conversion and re-conversion process, journalists use their own experience (frame) to separate events from the original context. On the other hand, it connects (or reframes) this event with other social meanings, creating new contextual meanings (Zang, 1999: 108–109).

Regarding the specific mechanism of the media frame, Entman states that essential factors are selection and salience: to frame is to select some aspects of a perceived reality and make them more salient in a communicating text, in such a way as to display meaning interpretation, causal inference, moral evaluation, and/or treatment recommendation. It is summarized by Zang Guoren as selection mechanism and reorganization mechanism. Just as Goffman said, the real fragments are often reshaped by "a pair of invisible hands", thus producing a result that conforms to the original structure but has different contents. The "invisible hands" often select real parts and rearrange and combine them to achieve the "reproduction" or conversion into the real (Zang, 1999: 45–50).

The audience frame means that when the audience contacts the media, their schemata will influence the selection, interpretation and judgment of contents of news report. That is to say, in the face of news media, the audience does not receive passively, but filters news content through their own cognitive frame, and interprets the authenticity, objectivity and humanistic care of news reports.

4.2.2 Relationship Between Media Frames and Audience Frames

How do the media frame and audience frame work?

First, let's look at the media frame (see Fig. 4.5). Zang Guoren thinks that the connotation of the media frame includes three parts: a media institutional frame, a personal frame, and a text frame. The media institutional frame refers to a series of routines and procedures formulated in journalism, which determine whether social events will be selected and reported. It is a framework mechanism for the media. The personal frame of journalism includes two parts: journalists and editors. That means journalists are influenced by their own cognitive structure. They have a set of ethnomethodology theory. Based on it they formulate their work goals, but they are subject to cognitive structures and cannot avoid self-stereotypes or prejudices. The text frame refers to the process of constructing a linguistic meaning in news writing text. Language and other symbolic information are the transformation of social reality. In this process, the text has a frame itself (Zang, 1999: 109–148). The media frame is actually the result of three sub-frames. The process of the final formation of news reports in the objective world is—objective truth to media institutional frame to journalist frame to text frame to editor's personal frame to news report.

Second, let's examine the audience frame. The audience frame is a filter-like selection for the content of news reports. It does not choose content that is completely incompatible with the frame requirements. Instead it assimilates or transforms other content consistent with its own frame. The content that is inconsistent with its own frame will be changed. The audience frame has a meaningful interpretation of the report, and affects behaviors in psychological level and media consumption. These actions in turn affect media frame.

Thus, the news is not really formed after media coverage, but after media frame and audience frame work together. In other words, a piece of news actually has meaning after the interaction between the media frame and the audience frame. The interactions of a series of media frame and audience frame constitute the entire

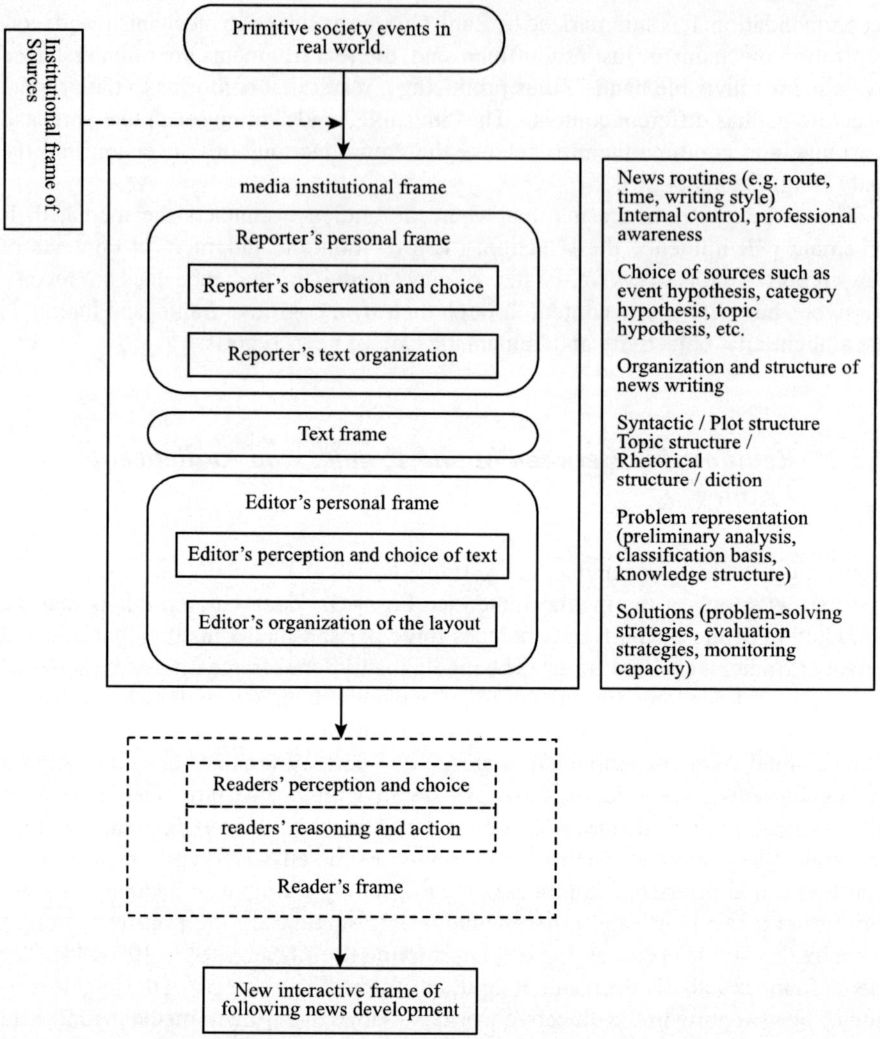

Fig. 4.5 The structure for basic connotation of news (media) frame

communication process. Woo (1994) and other scholars believe that the frame is the result of interaction between journalists, sources, audiences and social situations (Zang, 1999: 113).

4.2.3 Generation of Media Credibility: Duplication of Media Frame and Audience Frame

The interactive relationship between the media frame and the audience frame is actually a kind of game. To continue this game, one condition to be met is to establish a certain trust relationship between the media frame and the audience frame. From the perspective of the game theory of trust, if initial media credibility is generated in the initial game between the media frame and the audience frame, this level of media credibility will enable subsequent games to occur. Therefore, media credibility is established at different stages with varying degrees. In other words, media credibility is generated in repeated games between the media frame and the audience frame. If the game between the media frame and the audience frame cannot occur, or it can occur only once, then credibility will not be built.

(1) *The internal game mechanism of the media frame and the audience frame.*

Social events change from objective truth to truth in media after news (media) framing. After the media framing, it is impossible for the media to be a true copy of reality. News reports are, as Lippman says, like "the beam of a searchlight that moves restlessly about, bringing one episode and then another out of darkness into vision (Zang, 1999: 68). There is a difference between the truth in media and the objective truth. In the process of communication, the authenticity of the media has become the truth received by the audience after the audience framing. In this way there is a relationship of objective truth, media truth, and audience truth.

The audience's evaluation dimensions of media credibility such as objectiveness and impartiality are not comparisons between media authenticity and objective truth. Its mechanism is to contrast between media authenticity and audience authenticity, which is the audience frame's evaluation of media authenticity. The audience frame evaluates media authenticity at two levels. First, at the fact level the audience frame judges whether the media is authentic at the basic level through the audience frame's interpretation of the objective truth, based on the audience's perception of the media authenticity. Second, at the value level, the audience frame judges the basic value of truth in the media (see Fig. 4.6).

In order to obtain the judgment of the audience's trust, it is necessary for the media to constantly adjust its own frame to match requirements of the audience frame as much as possible so that the audience can follow up with the media contacts. Meanwhile, the majority of media more or less have their own news ideas, that is, the media frame is not always to meet various requirements of the audience frame, but to adjust to meet requirements of the audience frame based on its own frame positioning. In the repeated game between the media frame and the audience frame, media credibility is built.

(2) *The game revenue of the media frame and the audience frame.*

Credibility is generated through repeated games between the media frame and the audience frame. Then, what are the benefits of both parties in the game? For the

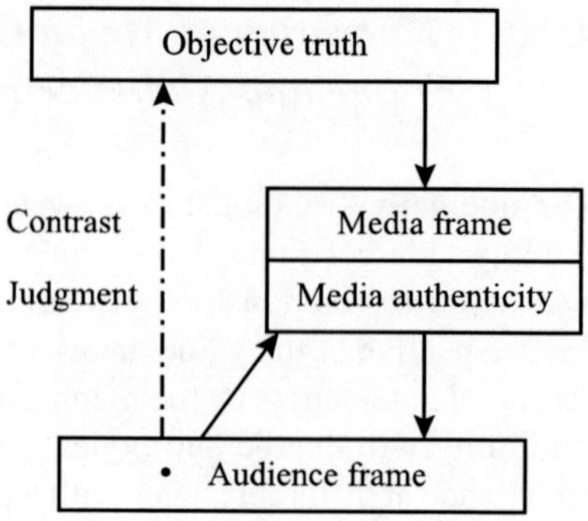

Fig. 4.6 The audience frame's evaluation of media authenticity

audience, the public in different regions are different. As discussed in the analysis of judgment dimensions of credibility earlier in this book, in different political systems, as the media plays different functions, the audience has different expectations for the media and the judgment dimensions of credibility are different. The audience's benefit is the perception of the specific content of these judgment dimensions. For example, in China, the judgment dimension of media credibility is authority and availability. For the audience, their benefits from the game is whether the information they have access to are authoritative or useful. As for the media, the game revenue of the Party newspaper and the Party journal is whether they play a propaganda role or a guide for public opinion. In addition to the propaganda and guide for public opinion, market-oriented media depend on whether economic benefits have been obtained. For the American media, the audience's benefit from the game is whether news information they are exposed to is factual and accurate. The gain of the media is to have a larger audience, to obtain economic benefits, or to seek greater political benefits.

If the audience frame does not have game gains during the game with the media frame, then the game cannot proceed. Conversely, it is the gain of the game which contributes to the next game.

4.3 Development Phases of Media Credibility

Media credibility is generated through the repeated game of the media frame and the audience frame. Media credibility, as trust, has a vertical development process. From the perspective of the three-stage model of Lewicki and Bunker, the development stages of media credibility are the calculus stage, the knowledge stage and the identification stage.

4.3.1 Rational Calculus Stage

This is the first stage of media credibility, which can be called calculus-based media credibility, or simply calculus-based credibility. There are three obvious characteristics at this stage.

First, the game between the audience frame and the media frame is a rational calculation. At this stage, since the audience cannot predict the performance of the media, they have to judge whether to trust the media by their perception after contacting the media. For the audience, the calculation of benefits may be done for each media contact. It is likely to wonder whether this newspaper is authoritative, whether the local news can be trusted or not, and whether complaining to it about the property has an effect. This stage is more common on newly-created media. For example, if a piece of newspaper is published, it is unlikely for readers to trust it at the very start. Only after reading it for a period of time will there be judgments about trusting it or not. A medium is supposed to develop new audience. Since new audience may not have contacted the medium before, they do not know whether its reports are trustworthy. Therefore, at the very beginning audience's evaluation of its credibility will be gradually made through contacting it.

Second, calculus-based media credibility is weak. For the media at this stage, every performance may affect media credibility. Since every contact with the media is a game of trust between the media frame and the audience frame, they are not consistent each time, which may affect the audience evaluation of media credibility. If the audience's evaluation of the media in the game is low, it will affect the next sub-game. When the continuous evaluation is low and reduces the credibility to a minimum, it may affect the subsequent game. This effect is even more pronounced for the mainstream media. Therefore, the media need to address readers' responses carefully. Once it is found that there exist factors that undermine credibility, it is very important to be able to remediate or manage the outcome in a timely manner.

Third, according to the research on trust conducted by sociologists, calculus-based trust is generally guaranteed by the rewards received and the punishment for destroying trust. For the media, credibility is an important factor to win the audience, especially for mainstream media such as *The New York Times* and *The Washington Post*, or elite media such as *Southern Weekly* in China. The credibility is essential to win the audience. It can be said that credibility is the lifeline of these media. Losing the trust of the audience means losing the basics of survival. For some relatively popular media, credibility is also a factor that affects the size of the audience.

If the media cannot establish the credibility, the first penalty it receives may be an effect on the size of the audience, which is the basis for media markets. A decrease in the size of the audience means that the media's attraction to commercial and its other business capabilities are weakening and ultimately affect the overall economic gain. If the media establish due credibility, the rewards may be a positive effect on the size of the audience, on attracting more commercials, and ultimately on getting more economic gain.

Thus, at this stage the media frame needs to be actively adjusted to maintain consistency with the audience frame.

4.3.2 Knowledge Stage Between Rational and Emotional

The second stage is knowledge-based media credibility, which can also be called knowledge-based credibility. This stage is based on the audience's previous perception of the media. The audience trust in the media does not need to be judged after contact. It has certain expectations for future behavior of the media and can predict the next performance of the media. Knowledge-based credibility does not have the uncertainty in calculus-based credibility. The maintaining elements in knowledge-based credibility are predictable. However, due to its uncertainty, calculus-based credibility can only be maintained through the punishment mechanisms. For the audience, predictability costs less and is guaranteed.

According to Shapiro, the better the one knows the other party, the more accurately the one can predict what the other will do (Lewicki et al., 2003: 112). For the calculus-based credibility, the audience needs to make a judgment of trust after contacting the media. The important point is that the audience does not understand the media, that is, the information between the media and the audience is asymmetric. For the knowledge-based media credibility, the audience has some information of the media, and the media can meet the requirements of the audience. The information between them is relatively symmetrical.

This trust is best illustrated in surveys at newsstands and retail outlets. Here, let's look at the survey data of a group of Beijing newsstands and retail outlets in 2002 (see Table 4.1).

Regarding the first case of buying newspapers, "always buy certain newspapers more fixedly", the author made special observations and conducted interviews at newsstands in Chengdu, Chongqing and Beijing. Readers who buy certain newspapers frequently, usually go to the newsstand to buy newspapers by name. They don't need to read the front page or guide, and they are rarely affected by other newspapers' promotion activities at the newsstand. This was confirmed by interviews with newsstand owners. Most of them thought that these readers had formed a kind of habit. When they arrived at the newsstand, they paid directly and got the newspaper

Table 4.1 Buying newspapers randomly or fixedly (Unit: %)

Purchase type	Selection ratio
Always buy certain newspapers more fixedly	45.2
Not fixedly, randomly choose to buy	22.9
In two cases, there are more fixed purchases	24.8
In two cases, there are more random purchases	7.0

Source Yu (2003: 77)

by name without reading its introduction at all and caring about other newspapers. Sometimes the newsstands promoted other newspapers and had no obvious effect.

From the development stage of credibility, it can be explained that these newspapers have established a knowledge-based credibility for readers who buy the fixed newspapers. The readers already have an expectation for what the newspaper covers, and they do not need to choose, for they can predict the contents of newspaper that they like.

Compared with the weak credibility of calculus-based media credibility, the knowledge-based media credibility is more stable. A mutual trust has been established between the media and the audience, and will not be affected by a certain inconsistency between the media frame and the audience frame. In media consumption the audience will be more stable in contact with certain media, such as the readers fixedly purchasing newspapers found in the above survey of the newsstands and retail outlets.

Shapiro argues that the greater the predictability, the more trust can be strengthened. And the more accurate the prediction is, the more contact is required, that is, like calculus-based trust, the process of understanding is an iterative interaction (Koehn, 2003: 112–113). The same is true of media credibility. After the audience has established knowledge-based trust in the media, the more contact the audience has with the media, the clearer their expectations of the media, and the more predictable the behavior of the media. As a result, media credibility will be enhanced.

Regarding this proposition, we quoted a survey data made by a reader of *Chengdu Business Daily* in August 2002 to see the specific situation in China. The reader's evaluation of the newspaper credibility was based on the reader's frequency of reading the newspaper weekly, the time in reading each newspaper, and the proportion of reading the newspaper. The analysis found that the Pearson coefficient represented the relationship between them. Among them, the variable with the proportion of reading the newspaper content has the highest correlation, the Pearson coefficient is 0.573, that is, the more the content of the newspaper reading, the more credible the newspaper is. Correlation with frequency of reading newspapers per week follows, Pearson coefficient was 0.25. It had the lowest correlation with the time spent reading each newspaper, with a Pearson coefficient of 0.197, which was relatively weak.

4.3.3 Emotional Identification Stage

The identification-based media credibility is the highest level, which is also called the identification media credibility or the identification credibility. At this level, the audience highly identifies the media, and treats it as their own. They can share the success of the media, and actively think for it and even act for it. From the perspective of frame theory, the audience frame and the media frame have a certain degree of isomorphism.

Some hundred-year-old newspapers and Chinese Party newspapers and journals are typical media with a credible identity. For example, a hundred-year-old newspaper like *The New York Times* has a group of readers who strongly agree with the style and philosophy of the newspaper due to the accumulation of history. They trust the newspaper since reports of the newspaper conform to their cognitive structure. Reading newspapers is not only to obtain temporary gains (such as knowing some advertisements, current affairs), but also to gain a deeper psychological identity. Despite the revelations of fraud like Blair at *The New York Times* may affect a person's evaluation of the newspaper credibility. But for readers who agree with the values of the newspaper, it is difficult to affect their trust in the newspaper.

For the Party newspapers and journals in China such as *People's Daily* and *Qiushi Journal.* Many old revolutionaries and Party members actually internalize their frames in these media frames, and do their best to maintain the credibility of these media and support the concept of the press frame.

For the media, the credibility at this level is the highest return. First, the higher the degree of identification-based credibility, the more stable the audience of the media will be. It won't happen that the audience don't trust the media or contact them anymore because of some incidents that damage media credibility. On the contrary, the audience may actively give feedback and suggestions, and even participate in some actions of the media. Second, the higher the degree of identification-based credibility, the more obvious the media's dissemination effect. Because the media frame and the audience frame have a certain degree of isomorphism, their concepts are basically the same. The media reports are easily accepted by the audience.

The media have a high degree of identification and credibility, which does not mean they don't need to care about the relationship with the audience. The credibility at this stage needs to be maintained at any time. Some maintenance of calculus-based credibility and knowledge-based credibility works for identification-based credibility. Generally, identification-based credibility is not easy to break, but once broken, it is a breakdown of the relationship. The cost of repairing and restoring the relationship after its breaking is incomparable to the calculus-based credibility and the knowledge-based credibility. This breakdown is not simply a matter of psychological feeling, but a deep destruction of emotions and ideas.

4.3.4 The Development of Three-Level Media Credibility

There are several points about the development of the relationship among the three levels of calculus, knowledge, and identification media credibility.

(1) Judging from the development path of media credibility, it starts from calculus credibility, and the path is in turn calculus-knowledge-identification. The audience's trust in a medium is initially a calculus type. The audience needs to contact the medium to judge its trust level. For the media, their performance may have an effect on their credibility. When the relationship develops to a certain degree, the audience begins to have a certain knowledge of the media performance and has a certain expectation for their future performance (that is,

it can be predicted what the media will do next) At this time, media credibility reaches the knowledge stage. The audience can not only predict the performance of the media, but also agree with the media philosophy, and even be ready to think for the media and treat the media as their own. In this way they will gain the identification credibility.

(2) The process of calculus media credibility to knowledge media credibility to identification media credibility is one in which the audience perception of the media going from shallow to deep. Media credibility develops from calculus to knowledge and then to identification. The development of each level is actually a change in the cognitive structure of the audience, and a process of audience perception from shallow to deep. At the calculus stage, the audience is rational to the media. At the stage of identification credibility, the audience and the media have a certain degree of isomorphism in the cognitive framework, which is emotional.

(3) In one kind of the media, components of credibility are same at three stages, but their ratios in these three stages are different in different media. For example, for a well-known newspaper with high credibility, its components of credibility have identification, calculus and knowledge. The same type of newly-launched newspaper which is formal, there exist all three levels in credibility. It is assumed that their credibility is same, but the difference may be that the former has a larger proportion of identification credibility. The latter may have a larger proportion of calculus credibility.

4.4 Control Analysis of Media Credibility Generation

The control of media credibility is directed at the control of the media itself. We discussed factors that affect media credibility, including a macro social environment and a micro audience. These external factors are beyond the control of the media. What the media can control is its constitutes.

For the media, credibility is an important resource. As Professor Albert Hirschman states, trust, like other moral resources, grows with use and decays with disuse. These are resources whose supply may well increase rather than decrease through use; second, these resources do not remain intact if they stay unused; like the ability to speak a foreign language or to play the piano, these moral resources are likely to become depleted and to atrophy if not used (Dasgupta, 2000). In other words, the media needs to adjust itself to meet requirements of the generation of credibility so as to maintain or improve credibility. The question to be discussed here is: how does the media control the generation of credibility?

When previously discussing the generation of media credibility, it was believed that credibility was generated by the repeated game of the media frame and the audience frame. From the perspective of the frame theory, it is important to discuss how the media itself maintains the credibility through control. The sub-frame of the media frame includes three parts: a media institutional frame, a personal frame, and a text frame. The personal frame of journalism is divided into the personal frame of journalists and the personal frame of editors. They are discussed respectively in the following.

4.4.1 Control of the Media Institutional Frame

The media institutional frame takes the first step in the generation of media credibility. To find a way to improve the credibility of the media, we first analyze the case of the media institutional frame from two perspectives, internal and external appeals.

4.4.1.1 Internal Demands of the Media Institutional Frame

The internal demands of the media institutional frame include the rules and regulations of the media, the methods of internal control and professional awareness. Zang Guoren argues that the news institutional frame includes news routines, internal control and professional awareness. News routines refer to the practices and procedures formulated by news institutions, which can be regarded as the first mechanism of social events in the news media frame, such as determining what social events can be selected and reported. In the view of Zheng Ruicheng, the internal control refers to the working environment of news editors. There are three control methods: punishment, reward, and regulation. If the news is not written well or the big news is missed, the superior will blame or threaten to fire the staff. This is a punitive effect; When a reporter gets a scoop, the superior will give verbal praise or material compensation, which is a reward. Norms aim to convince people, and media institutions communicate their institutional culture through various internal channels. Professional awareness generally refers to how to obtain job qualifications, how to perceive the role of mass media, how to maintain the independence to free from external interference, how to properly interview, and how to cope with the journalism and corporate profits and address professional and entertainment styles (Zang, 1999: 114–122).

An important part of news routine is the positioning of news media. Once determined, it is not easy to change. Professional awareness has been valued in China, especially in the context of the marketization process in the past ten years, which has had a greater impact on the concept of pressmen. Education of news concept and journalist qualification tests have been carried out nationwide.

The focus here is on internal control. There have been many studies on the operation of internal control. Lo (1993: 12–16) thinks that to increase the credibility of the news media, the first thing is to ensure that news reports must be as accurate and fair as possible. This requires the press to strictly prohibit reporters from fabricating news, and requires reporters to make every effort to verify the truth and avoid publishing false news. One of the effective methods is to require reporters to point out a clear source of information in the news report. The other one is to adhere to the principle of separating facts from opinion in news reports. The second is to strengthen the moral concepts of journalists, such as refusing to accept gifts from parties and free travel, etc.

The commercialization of the media goes to the editing system, which is an important cause of problems in internal management. The internal control should be improved from this perspective. For example, after the Blair incident in *The New York Times*, Professor Al Giordano thought that *the Times* should reflect on its own operating model. After becoming as big as a dinosaur, *the Times*' internal management was disordered and flawed. At the same time, Al Giordano stated that many business press in the United States became "sweatshops" for young people. The investigating articles in *The New York Times* only showed "a huge tip of the iceberg on the ocean of business media." Giordano did a simple calculation, in which 73 articles were published in seven months, with an average of 10 articles per month and one article every three days. With five working days per week, in the strong editorial board of *The New York Times*, a new practitioner needs to publish an article on average two days. And he or she must travel frequently to write "real life stories".

Some market-oriented media in China begin to establish relevant internal management systems to pursue marketing goals. For example, the internal evaluation system of the media stipulates that journalists must complete a certain amount of work to have a basic income. The level of income is only related to "yield". As some people in the press think, in the current situation where the relationship between journalists and newspapers has actually become a relationship between workers and factories. More and more journalists are only using this profession as a means of livelihood." (Li, 2005).

Below is an excerpt from an interview with Harry Moses, the senior director of CBS *60 Minutes*, to see how this old American show guarantees the authenticity of its report (Zhu et al., 2004).

Reporter: How does American investigative reporting guarantee the authenticity of news? A news work is a product, and we must consider such questions as whether to control the production process of the interview, or whether some smart people with white hair will judge its authenticity based on experience? Does CBS have written regulations or process controls in this regard?

Harry Moses: It depends on the source of the story. If the gossip news from the tabloid says that a certain politician is a vampire, you have to check it again. But if it comes from a big press like The Washington Post, it has a sense of trust and is worthy of continued investigation. Of course, there will be some bad apples. Last year, The New York Times had a news fraud, and he was finally found. Despite the adverse effects, this kind of thing rarely happened.

If a person goes against the moral conscience and deliberately makes up the news, no matter what method is used to conceal it, he or she will be caught. For example, in "60 minutes", if someone makes up a story, it may take some time to be discovered. Therefore, when media companies hire people, they will be very careful, and they will definitely choose people who truly respect the value of news.

Reporter: Does the quality control and supervision exist in the American media? For example, interview notes, attendance records, source records, etc. Do you read the manuscripts directly or do comprehensive inspections?

> Harry Moses: CBS has its own standards and guidelines. An editor will examine the recordings (videos) of all interviews and review all news materials. The problem is that with these guidelines, it is impossible to record everything. If you are a journalist with a conscience, you can judge whether you are right or wrong. This is an intuition.

The specific control measures of the media institutional frame mentioned above are worthy of reference for any media. For the media in China, one thing that needs to be emphasized is the political attributes of the media. This is the primary media frame. As for how to make the media's political frame easily acceptable to the audience and build trust, they must have their own operation mode. For example, in the case of typical reports, does the media frame operate in the old way? or they report in a new way that meets audience expectations, which has much to do with whether the audience trusts the content of the report.

4.4.1.2 External Demands of the Media Institutional Frame

The external demands of the media institutional framework refer to the public relations of the media, that is, how the media handles the relationship with the public. This is an operational content and there are many practices. Establishing interaction with the audience is the most common type of media public relations behavior, such as readers' festivals organized by many urban newspapers in China in recent years, and establishing readers' clubs or readers' societies. It is also common to organize or participate in public welfare activities to establish a good image among the public. Some media make opportunities to promote their image and ideas.

4.4.2 Control of the Personal Frame

The personal frame of journalism refers to the cognitive structure of journalists. It is the psychological structure of journalists' personal choice, interpretation, or organization of external affairs. It mainly includes two parts, the reporter frame and the editor frame. Some scholars emphasize that news actually represents a component of social events in journalists and editors' personal frame after excluding other components (Zang, 1999: 123, 128–129). In other words, a journalist reports the objective truth through subjective processing. Journalists and editors are different individuals, and there are differences in personal cognitive framework. This difference in the personal frame of reporters and editors has effect on journalistic principles of the authenticity. There are two aspects of control. First, the pressmen are lack of professional awareness, professionalism and understanding of the audience. This is the reason for influence on media credibility. It is not caused by their wishes, unawareness or inability. Second, journalists are aware that media credibility can be damaged by their personal interests. Managing conflicts of interest is an important part in controlling the personal frame.

Lo Ven-hwei et al. found that receiving freebies and part-time jobs are the most serious ethical issues facing journalists from Taiwan of China. In other words, journalists must play their professional roles on the one hand, and face the temptation of commercial interests on the other hand. Conflicts of interests arise when a person plays two or more roles at the same time, or when performing professional tasks, facing conflicts between personal interests, institutional interests, and professional responsibilities. Wulfemeyer conducted empirical research on the conflict of interests in journalism. According to his analysis, potential conflicts of interests in journalism tend to fall into five categories: (1) freebies and junkets; (2) moonlighting; (3) involvements with communities; (4) personal attitudes, beliefs, values and socioeconomic status, and (5) external pressures from owners, bosses, advertisers, government (Lo et al., 2001).

Sanders and Chang (1977), two professors at the University of Missouri School of Journalism, in a survey of 123 newspaper editors, asked them to evaluate whether freebies would affect the objectivity of news reporting. It was found that about 77% of the respondents thought acceptance of free trip to foreign countries would affect most journalist's objectivity in reporting related stories. About 71% felt that acceptance of free memberships to country clubs would affect most journalist's objectivity. More than half of the respondents believed that acceptance of freebies, free domestic trips, discount at golf courses, and free season tickets for college or professional sports would affect the objectivity and integrity of most journalist (Lo et al., 2001).

In addition to economic interests, journalists may have interests in social relations and political appeals and may face conflicts in their professional roles. In this way their cognitive frame will deviate objectivity and impartiality required by professional roles. News operations under this conflict framework will inevitably be reflected in news reports, and media credibility will ultimately be affected.

For the control of these new issues in the news personal frame, here is an excerpt from the introduction of Harry Moses, a senior director of CBS 60 minutes at an academic conference, to see the control method of this old show (Zhu et al., 2004).

Reporter: What is the average age of a reporter for 60 minutes?

Harry Moses: Around 47 or 48 years old, almost 60 years old on average in the past.

Reporter: When you do surveys related to the public interests, what methods can you take to achieve relatively perfect results? What are the essential qualities of a good investigative reporter?

Louis Wiley: There are many local newspapers in the United States with dedicated news investigation teams. Different reporters have different interview techniques. Some can understand police records, while others can understand hospital documents and financial information within the company. For some more difficult topics, special training is conducted in the newspaper. There are protections for informants and relationship maintenance. Good investigative journalists need to have good qualities meet with different people and persuade them to provide information. Some journalists have this talent.

For specific interview techniques, there are different answers: For example, although the source cannot be disclosed in the report, you can still publish relevant information. In this case, you need at least two sources. This is the minimum requirement. If doing a TV report, you can shoot a shot that knocks on the door but no one opens; In print media reports, you can directly describe the other party's refusal to be interviewed. You can explain in the published report that you still have interest and expectation in the response of those who are unwilling to be interviewed, and will continue to pay attention. This method may convince the other party to agree to answer your questions after reading the report.

From the interview with Harry Moses, the director of CBS *60 Minutes*, it can be found that there are several ways to control the personal frame of the media. The first is journalists' age and experience; second, the training for journalists; third, the expertise of journalists; fourth, the way to determine the message; and the fifth is the way of interviewing.

For news pressmen in China, there are two more points worth noting: one is age, and the other is that the political attributes of the media requires journalists with a certain political quality. The age requirement of reporters in CBS *60 minutes* is different from that of some highly market-oriented media in China. Whether it is television, newspapers, or journals, whether it is urban newspapers, or political and economic media that follow the mainstream line, the age of journalists required by these media in China is generally young. To report on a complex social event and grasp its objective nature, one's age and experience are extremely important. Simply using young age as a key indicator for selecting journalists is important. For large mainstream newspapers or media, it is worth valuing the requirements of *60 minutes* of CBS.

4.4.3 Text Frame Control

The news text frame is to organize the story into a meaningful whole through the use of syntactic structure, plot structure, subject structure, rhetorical structure, etc. The choice and combination of words in a news story is not a small trick, but an important tool to determine the context of the debate, the definition of issues, and to trigger the reader's response and start discussions (Zang, 1999: 143).

Voice is an important content of news text frame control. In short, it is the narrative way to report news, and arrange the text structure in a way that the audience likes to accept and that impresses the audience. This requires the media to have a more accurate understanding of the receptive mind of the audience.

The impact of the text frame on the audience can be found in this data from Taiwan of China. Zhu Yufen tested the reader's emotional intensity with news reports in traditional inverted pyramid format and new reports in common pyramid format, and found that the former is less likely to cause readers to be emotionally involved. Even in the introduction, although this writing structure was supposed to be the most important paragraph of a news story, it could not arouse readers' interest; Comparatively, writing in narrative form can trigger readers' strong emotional response (Zang, 1999: 147).

Zhu Yufen conducted this research in Taiwan of China. There needs some empirical data for its impact on China' mainland or Western countries and regions. But it is clear that the structure of the text itself is a frame for objective truth. The same thing, expressed in different words, or different rhetoric, will have different moods, and have different perceptions for readers. The same is true of television. Different voices produce different cognitive effects.

Different text frames have different effects on media credibility. In December 2004, a "pre-trial survey of media credibility research" conducted by the Institute of Public Opinion Research at Renmin University of China non-randomly surveyed 738 residents in the Haidian District of Beijing and asked them to evaluate the credibility for *CCTV*, *Beijing TV*, *Hunan Satellite TV*, *Dragon TV* and *Phoenix Satellite TV*. Among the five televisions *Phoenix Satellite TV* got the highest score in the credibility evaluation of their viewers. In the research literature of *Phoenix Satellite TV*, one of the most obvious features is its different voice compared to other TV stations. *Phoenix Satellite TV* is basically the same in ideology as the other TV in China's mainland, but same news content is reported in different voices, and the audience can easily distinguish the differences between the two because the differences in voices give the audience different perceptions. This differences in voice affect the audience perception of trust.

In short, for the media institutional frame, the personal frame and the text frame, the control of credibility is a holistic task, which requires various parts working together. This is like "the wooden barrel", whose capacity is determined not by the longest wooden bars, but by the shortest. The establishment of media credibility does not depend on the interior of the media only. For the media, it is necessary to manage the audience expectations for the media and the changes in the communication environment (such as dimensions to judge whether the media is credible, and whether the media's voice meets the audience's receptive requirements, etc.). Managing from the perspective of the relationship with the audience can be more effective.

Chapter 5
Impact of Credibility on Media Functions

In the early stage of SARS in the spring of 2003, the absence of the media caused panic among the people, and media credibility became one of the hot topics of discussion in our academic circles. The topic discussed more frequently is that the decrease of media credibility would affect the communication effect of the media. American communication scholar Jacobson (1969: 20) believes that media credibility directly affects the effect of communication. On the theoretical level source credibility is significant to communication research. In terms of changing the audience's attitudes, high credibility sources are more effective than low credibility sources from the perspective of cognitive response (Stamm & Bube, 1994: 105–123). Research reports from the Roper Organization and the American Society of Newspaper Editors show that the credibility of American newspapers has been declining for the past 50 years, especially after the disclosure of the fraud committed by Blair, the young journalist of *The New York Times*. Media credibility has been a common concern.

What impact does the credibility have on the function of media? A clear picture of media functions is shown in this chapter.

5.1 Analysis of Media Functions

5.1.1 Three Perspectives of Media Functions

5.1.1.1 Define the Relationship Between "Function" and "Effect" Which Are Frequently Mentioned in Communication

As for the result of media activities, from the perspective of the media itself, it is "effect", which is usually called "communication effect"; while it is "function" from the perspective of society or the audience. They are two sides of the same issue, and the content of their expression is the same. For example, Maozheng (1994: 329) believes that those who study mass communication are often bewildered by

G. Yu, *Research on the Communication Effects and Mass Media Credibility in China*,
https://doi.org/10.1007/978-981-19-6242-4_5

the terms of "effect" and "function". Some concepts such as "to socialize" or "to focus attention" of the media are classified as "effect", while others are classified as "function". In fact, "function" and "effect" only examine the media activities from different angles. The results of the media are "effect" from the perspective of the medium itself, and "function" from the perspective of the society or the viewer. Many differences come from different angles. In mass communication activities or communication research, it is gradually viewer-based and individual-centered in the society that relatively improve the research of human value. After expanding research on the relationship between social groups and individuals, it becomes a focus on function. On the other hand, "effect" refers to the final result of media activities, while "function" values the impact of its "process". In this study, the term "function" is used to discuss research on "media function".

5.1.1.2 Distinction of the Research Perspectives for Media Functions

The scholars in China examine functions of mass media from three levels: (1) the inherent social function of mass media itself. This is determined by the communication characteristics of the media. (2) The owners of mass media and all parties in society have the expectations of media functions, which is brought about by the instrumental nature of the media. For example, the politics circle takes it as a weapon for their political work, while businesses hold it as a means of doing business, etc. Strictly speaking, this level should be regarded as "the view of media function". (3) The actual impact and role of mass media on the individual and the audience. This level is often called "media effect" (Guoliang, 2001: 59).

Liangrong (1997: 85) argues that functions of mass media can be summarized into three representative views. The first view is that function refers to the role of mass media in its interaction with society. The second point of view is that function refers to the role that the host of the news media and all parties of society wish (or expect, require) the news media to play for their own interests, or a tool that the news media is expected for. The third point of view is that function is the role and influence of mass media in society during the communication process.

The meanings of the above three statements are the same. They are viewed from the perspectives of society, the public and the media. The first is to examine media activities from a social perspective, that is, the social function of the media. The second is to view the function of the media from a public perspective, that is, a "tool" function of the media. The third is from the perspective of the media, that is, the effect of the media communication.

5.1.1.3 Further Analysis of Media Functions from the Above Three Perspectives

Regarding the specific content of mass media functions, scholars have made more discussions from the perspective of "effect" and "function". There has been no consensus since functions of media are changing with the development of society. On the other hand, academic research is developing, and the understanding of media functions is deepening. It is impossible to reach a final conclusion.

(1) From the social and public perspective, concerning the media functions, Lasswell, one of the four pioneers of the mass communications, proposes three major social functions in the article "The Structure and Function of Communication in Society": (a) the surveillance of the environment; (b) the correlation of the parts of society in responding to the environment; (c) the transmission of the social heritage from one generation to the next. Later, Wright, a famous sociologist supplemented Lasswell's theory of three functions in his book *Mass Communication* and added the fourth function of "entertainment".

Schramm divides media functions from a macro-social level and makes a general summary of media functions. At the macro level, the functions of mass media are classified into three categories: economic, political and social (see Table 5.1).

(2) From the perspective of the medium itself, the function of the media is the effect of communication. The study of communication effect is a research focus of the empirical school. Starting from the early and middle twentieth century, the theory of communication effect evolved from the "bullet theory" of the strong effect theory to the limited effect theory and to macro-level effect theory in the 1970s. The main research on communication effect involves the agenda setting theory, the spiral of silence theory, the cultivation theory and the knowledge gap theory.

Table 5.1 Schramm's summary of the social function of mass communication

Political function	Economic function	General social function
1. Surveillance (collecting information) 2. Correlation (interpreting information; developing, disseminating and implementing policies) 3. Transmission of social heritage, laws and customs	1. Information on resources and opportunities to buy and sell 2. Interpreting this information; formulating economic policies; activating and managing markets 3. Creating economic activities	1. Information about social norms, roles, etc.; accepting or rejecting them 2. Correlating public perception and willingness; exercising social control 3. Transmitting social norms and roles to new members of society 4. Entertainment (recreational activities, getting rid of work and real-life problems, incidental learning and socialization)

Source Qingguang (1999: 114–115)

5.1.2 The Two Levels of Analysis Framework for Media Functions

This book analyzes media functions in two levels.

On the first level, we examine the media functions from the perspective of social systems theory. In the previous analysis, the social systems are classified into three subsystems–cultural, political, and economic. The media functions on the first subsystem level of the social systems are the influences of the media on the three social subsystems, that is, the media functions are presented in three aspects: political function, economic function and cultural function.

On the second level, the specific content of the political, economic, and cultural functions of the media are further observed. From the perspectives of society, the public and the media, the contents of the general social function, "tool" function and the communication effect are included respectively into three social subsystems—political, economic and cultural. In the subsystem political functions cover specific content such as agenda setting and coordination (political communication), economic functions including economic income and other content, and cultural functions including cultural heritage (teacher's roles), shaping and entertainment.

5.2 Impact of Credibility on Political Functions of the Media

5.2.1 The Political Function of the Media

The political function of mass media refers to the specific role played by the media in the political field. In the eyes of scholars from the radical Frankfurt school, the media not only act as a government's "microphone" and a tool of power but also a tool used by the government to maintain ideology and spread the will of the ruling class. It is even an ideology that directly plays an ideological role in social control and maintains the legitimacy of government. As Horkheimer and Adorno mentioned, the broadcasting system is a private enterprise, but it represents the power of the whole country. Chester Farm is nothing but a country's tobacco supply, but radio broadcasting is the nation's microphone. Marcuse warned, can one really distinguish between mass media as instruments of information and entertainment, and as agents of manipulation and indoctrination? It must be remembered that mass media at first glance is a tool for disseminating information and providing entertainment, but mass media that do not substantially function as ideological guidance and political control do not exist in modern society. As long as it is the media, even the entertainment media, it has to play a certain political role.

As discussed in this chapter, there are two perspectives on the specific content of the political function of mass media: the social perspective and the media perspective.

The first is from the social perspective, Schramm summarized the political functions of the media into three points: surveillance (collecting information), correlation (interpreting information, formulating, disseminating and implementing policies), and transmission of social heritage, laws and customs.

The second is from the perspective of the medium itself. In contrast to Schramm's social perspective, political scholars have concretely summarized the political function of the media from the perspective of the medium itself. Some scholars of journalism and communication have cited these statements of political scholars. For example, according to the literature of political scholars, journalists summarize the political functions of mass media into six aspects: political participation, agenda-setting, public opinion supervision, political communication, political control, and political socialization (Kun, 2003: 140).

In addition, the core of the media in China is positioned as its "mouthpiece" function. From this perspective, the relationship between the Party, the government and the media is intertwined, which has been discussed above. When a piece of news is reported on the TV or newspaper, it means that the government approves it and it represents the voice of the Party and the government. In other words, the media is an "organ" of the Party and the government, and an extension of the government functions. Its image is closely related to the image of the government. The credibility of the media affects the credibility of the government. Therefore, there is another political function of mass media, i.e. the function of representing the image of the government.

The following is to analyze the political functions of the media from the second perspective. The specific content of the political functions of mass media can be summarized as political participation, agenda-setting, public opinion supervision, political communication, political control, political socialization and government image. The following examines the impacts of political credibility on the above seven political functions respectively.

5.2.2 The Impact of Credibility on Agenda-Setting

5.2.2.1 Agenda-Setting

Regarding agenda-setting, Cohen argues that the newspapers and reviews cannot tell their readers what to think about, but they are effective at making readers think. Empirical studies conducted by Maxwell McCombs and Donald Shaw in 1968 and 1972 respectively confirmed this hypothesis. It means that mass media influences the public's priorities by funneling attention to certain issues while ignoring others. Due to the media's deliberate neglect, issues that should have received attention would be ignored by the public, which could affect the public opinion. And for the issues that mass media focus on, the degree of attention paid by the public to the importance of these issues is consistent with the sequential order established by mass media for them.

In the research on the connotation of agenda setting, scholars have made a lot of empirical discussions on factors influencing media and audience, scheduling, procedures and orientation in the "media agenda" and "public agenda" paradigms. On the development of agenda-setting extension, the research focuses on who set the media agenda and the causal relationship between the media agenda and the public agenda. Westley believes that pressure groups or special interest groups can make an issue become a media agenda. Funkhouser proposed five ways to get the attention of the media and conducted a comparative study of media agendas (Hongzhong, 2001).

The role of agenda setting in politics is analyzed from two perspectives. The first is the setting of political issues in the paradigm of "media agenda" and "public agenda", that is, how mass media makes a political issue or related one become public issues, i.e. playing a role in guiding public opinion. The concept of "opinion-orientation" first makes people think about the agenda setting, for it directly discusses how the media guides public opinion or changes the existing public opinion (Lidan, 1999: 78). The second is the agenda building in the opposite direction, from public agenda to media agenda, and then to policy agenda, that is, how a public agenda becomes the media agenda and affects the implementation and formulation of government policies.

In summary, the influence of media credibility on the political functions of agenda setting is mainly manifested in two aspects, the influence on the guidance of media opinion and on policy agenda.

5.2.2.2 The Influence of Credibility on Public Opinions

A prerequisite for political rule is that the government is able to guide public opinion and make public opinion follow a politically designed track. As some scholars think, unanimous opinions of the government and the public are manifested in the fusion of upper-level and lower-level opinions. This assimilation of public opinion is a prerequisite for the sustainable development of society. If the social system is trapped in the quagmire of decay and corruption, and the state machinery cannot effectively get rid of it, the lower-level opinions will directly aim at the upper-level opinions and cause opposition to the public opinion. According to the internal requirements of the social operating mechanism, it continues to manifest the course of nature and the will of the people (Jianming, 1998: 10). The most important way to guide public opinion is the agenda setting of mass media. Specifically, the government can set the media agenda, and make it affect the public agenda, so that it can guide the public opinion effectively. The guidance of public opinion is examined from the domestic and foreign environments.

(1) Investigate the impact of media credibility on agenda setting. American communication scholars Wanta and Hu (1994: 90–98) used path analysis to study the relationship between three audience attributes and agenda-setting effects in the agenda-setting process. The three attributes include the individuals' perceived

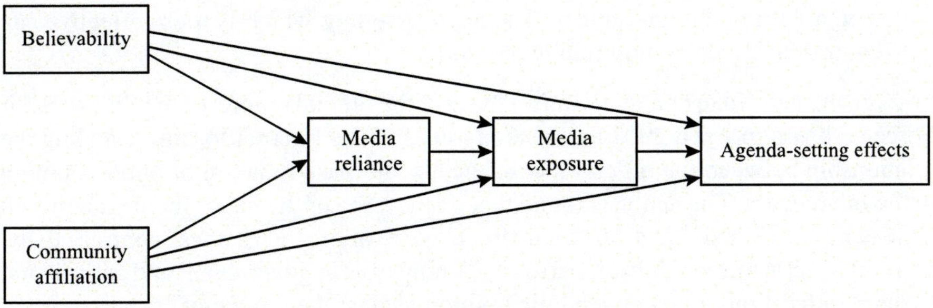

Fig. 5.1 Proposed path analysis model of agenda-setting effects

credibility of the news media, their reliance on the news media for information and their exposure to media messages. The model is shown in Fig. 5.1.

Wanta and Hu decomposed media credibility into two index dimensions: believability and community affiliation. They assume that if individuals perceive the media to be highly credible, they will rely on the media for information, and increase their exposure to media messages, and thus the agenda items set by the media will be more likely to have effects. The path analysis supports the model. The coefficients in the "believability" and "community affiliation" are statistically significant. Media credibility will lead to media reliance, which in turn will lead to media exposure, and then will result in agenda-setting effects.

Specifically, if individuals do not believe in a certain medium, it is difficult for them to be interested in the topics set by that medium. For instance, a Wall Street lawyer may regard the National Enquirer as less credible regarding international news than *The New York Times*; so when the lawyer reads a headline in the Enquirer about a new Soviet disarmament proposal, the medium's salience for this news item likely will not be accepted. The individual is informed about the news item by the media, but is not convinced that the item is important (Rogers & Dearing, 2000: 83).

(2) For the public opinion facing the domestic communication environment, high media credibility plays a positive role in guiding the public opinion, while low media credibility has little or no effect, even makes a negative or opposite effect.
(3) In an international communication context, media credibility is facing pressure of competition. The guiding role of the media's public opinion must be fully played. Media credibility is an essential prerequisite. With the advance of communication technology, the global village becomes a reality, and the channels for people to receive information are becoming diversified. Not only can they use their domestic communication channels, they can have easy access to information from foreign communication channels. In this context, at the national level the agenda setting of the media has become an increasingly important "soft power" in the competition between countries, that is, setting the media agenda can influence the public opinion of other countries, and even interfere with other countries' policies. In order to achieve the agenda-setting effects, media credibility becomes the core of the competition. The public believe in the

media with higher credibility. The agenda-setting which is more effective can contribute to controlling public opinion.

Xiguang and Xuan conducted an analysis of the official reports of the Sino-US Air Force collision in April 2001 and related topics of the Internet forum, revealing the relationship between the official media agenda and the formation of public opinion in the Internet era. The analysis found that on the issue of the major incident between China and the United States, the mainstream voice of the Internet was contrary to the voice of the government, which led to a declining trust in government and contributed to the proliferation of negative public opinion against the government.

Li and Qin (2001) concludes that from the perspective of international relations, if the public does not believe in the media of their country, he or she will turn to international media, or even media from hostile countries. Setting the agenda for other countries through the media and the Internet has become a kind of "soft power" invasion in international politics. Global and foreign media can influence a country's agenda-setting. The higher the public trust in media is, the more significant the effect of its agenda setting is. In the era of information globalization, it is impossible for our newspaper to expect its mainstream topics to continue attracting public attention. If China's mainstream media does not start reform early, acting according to the law of news, Western media will make agenda-setting for China.

5.2.2.3 The Influence of Credibility on Formation of Policy

Rogers and Dearing (2000: 67) believes that agenda research has two main research traditions that have been referred to as (1) agenda-setting, a process through which mass media communicate the relative importance of various issues and events to the public (an approach mainly pursued by mass communication researchers), and (2) agenda-building, a process through which the policy agendas of political elites are influenced by a variety of factors, including media agendas and public agendas (as shown in Fig. 5.2).

One defect of Rogers and Dearing's model is that it is one-way from the media agenda to the public agenda, and it is difficult to accurately define the inherent meaning of the agenda. In reality, the impact of the public agenda on the policy agenda is not directly related, but through the formation of the media agenda, and then affect the policy agenda. The relationship between the three is interactive. In other words, their sequential order is public agenda to media agenda to policy agenda.

Here, the media serves as a bridge. Media credibility has a direct effect on agenda-building. For example, in the United Kingdom, *The Sun* is a tabloid, and its credibility cannot be compared with *The Times*. When *The Sun* and *The Times* address an issue of public concern respectively, there is no doubt that the agenda setting of *The Times* has more effects on the policy agenda.

The credibility of mass media in China is high. Generally, the agenda-building has better effects. For instance, People's livelihood issues presented in general newspapers can be actively responded by the government. A typical case is the Sun Zhigang

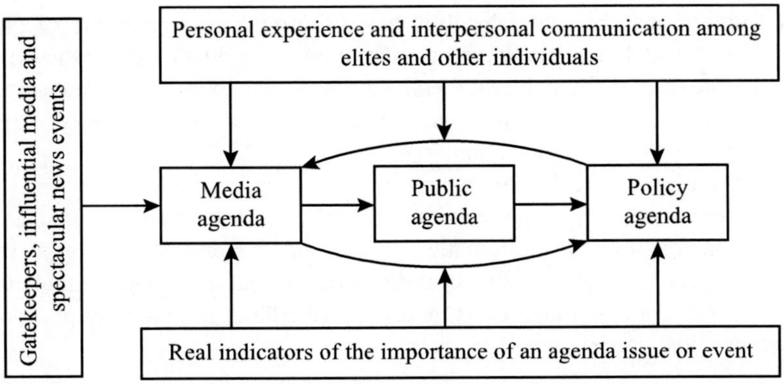

Fig. 5.2 Three major components of the agenda-setting process: media agenda, public agenda, and policy agenda.

incident in 2003. When *the Southern Metropolis Daily* reported the death of a young man named Sun Zhigang while in police custody sparked a public outcry. A big public agenda was formed on the Internet, which in turn formed a media agenda, and finally formed a policy agenda. The government complied with public opinion, and dismantled a controversial form of administrative detention called "Measures of Custody and Repatriation of Urban Vagrants and Beggars" implemented since May 1982. On June 20, 2003, Former Premier Wen Jiabao signed the decree of the State Council of the People's Republic of China, proclaiming "Measures on Aid and Management of Urban Vagrants and Beggars".

5.2.3 The Impact of Credibility on Political Participation

5.2.3.1 Political Participation of the Media

Political participation is the behavior of ordinary citizens participating in political life through various legal ways and affecting the composition, operation, rules, and policy of the political system. Citizens' political participation is of great significance to modern democratic systems. It builds legal supervision and restrictions on the government, such as the legal exercise of public power. Political participation is the basic right that citizens should enjoy.

For political participation to be effectively implemented, mass media as a social watchtower is an important means. Citizens' political participation includes political voting, political association, political expression, political contact and political apathy. In today's information society, political expression, political contact and political apathy are all closely related to the media, and even the media plays an important role. Specifically, for China, policies can only be effectively implemented on the premise that the general public has a thorough understanding of government affairs and policies. The government's policies, regulations, and the officials' behaviors must make the public understand that newspapers, radio, television and magazines are the fastest and most widespread means of communication, and they are important communication channels.

For the public, political participation through mass media includes two mutual directions: one is to learn about relevant political information such as policies from the media, and the political participation of mass media manifests itself as an important platform for the dissemination of political information. Another is that citizens express their opinions through mass media. The media is an important platform for citizens' political participation.

5.2.3.2 The Influence of Media Credibility on the Public Receiving Political Messages

Some scholars have pointed out from the perspective of the public domain that the quality of contemporary mass media should provide visibility. For the discussion of public affairs, although the media cannot really provide a space for interpersonal dialogue, it can greatly increase the visibility of the event by reporting. For example, reports of social protests do not allow all participants to discuss with each other, but they did expose the opinions of representatives. In other words, the function of mass media should be to provide adequate "information". Based on the information the people can talk and discuss on their own and make reasonable judgments (JinHua, 1995).

The previous research by Wanta and Hu confirmed that media credibility affects the audience's reliance on the media, and the media reliance affects the media exposure. In other words, the credibility of mass media is a prerequisite for its political participation. Mass media in China is the "mouthpiece" of the Party. It emphasizes the political nature. It is necessary to make good use of mass media's function of political message dissemination so that the public can receive political messages through mass media. Media credibility is a key factor. The following two messages show the importance of our media in this regard.

Message one:

> Morning News (Chief reporter Cui Hong) The Beijing Municipal Government's Legal Affairs Office said yesterday that in addition to protecting confidential information in accordance with the law, all contents related to the vital interests of the public, including administrative decision-making items, grounds and results, should be disclosed to the public.
>
> *The Municipal Legislative Affairs Office convened a related meeting yesterday and issued the "Opinions on Improving the work of Administrative Information Disclosure by Law", requiring administrative departments at all levels throughout the city to establish a system of disclosure of administrative information in accordance with the law. Administrative information should be disclosed to the society frequently and regularly to fully protect the public's*

right to information, participation and supervision. Meanwhile, the disclosure of administrative information according to law will be gradually promoted from the administrative level to the decision-making area.

The meeting stated that the public contents should include the contents related to the vital interests of the citizens in the legal argumentation of major administrative decision-making, and various schemes, plans, systems and related supervision related to administration according to law. Relevant information in drafting regulations and rules, including public opinions. According to the laws and regulations and the requirements of the Municipal Party Committee and municipal government, all contents related to administration according to law should be exposed to the public.

The main channels for disclosing administrative information in accordance with the law include improving public participation in drafting regulations, draft rules, and regulatory documents. Drafts that are of major or vital interest to the public must be widely exposed to the public through press conferences, demonstrations, or released to the public. It is necessary to give response to the public opinions and to make explanations of reasons for receiving the opinions.

Through different channels such as government bulletins, newspapers, and government websites, various regulatory documents related to the public interest are published in a timely manner, and various conditions are provided for the public inspection, including sending various types of government regulations to the community, "one-stop" office, etc. and allowing citizens to have free access to them (Cui Hong 2004).

Message two:

(Reporter Pang Jingtao) The reporter learned from the press conference held by the municipal government yesterday that the Chengdu Government Information Disclosure were discussed and approved by the executive meeting of the municipal government and would be officially implemented on May 1. It is one of the supporting documents for standardized service-oriented government. After Beijing and Guangzhou, Chengdu was the third city to issue corresponding regulations on government information disclosure.

……

Channels of government information disclosure:

The first is the special issue of the government information and other newspapers and magazines issued publicly. The second is the Chengdu Public Information Network and other government websites on the Internet. The third is public media such as radio and television. The fourth is to hold a press conference. In addition, other channels that are accessible to the public can be used according to specific circumstances. The regulations also make it clear that government agencies should actively disclose government information to citizens, legal persons and other social organizations, and should not charge any fees for any reason (Jingtao, 2004).

The government requires the publication of policy information through mass media, which is a manifestation of socialist democracy. But for whether it can be effectively achieved, the issue of media credibility becomes significant.

If the people do not trust the media, they will not be interested in what is reported or they will not contact it. Even if they do, they may treat political issues as entertainment. For example, a survey conducted by the University of Hong Kong in 2000 showed that 38.8% of the respondents considered the media to be irresponsible, and only 17.9% considered the media to be responsible. One of the reasons was the reporting style with "paparazzi" culture, which is popular in the media. What's more, there occurred some fake news. These events greatly weakened the reputation and image of the Hong Kong media among the public. According to a survey released by the University of Hong Kong in September 2003, The public in Hong Kong, China only scored 5.48 on media credibility, which is just a passing score. In July of the same year, the media received only 5.84 points. However, the score in 1997 was 6.55, which showed that the public are gradually losing their trust in the media. No wonder people in the society have been jokingly saying, "As long as you read newspapers and magazines with the attitude of reading novels or stories, you don't need to be serious about the contents."

5.2.3.3 The Influence of Credibility on the Public Opinion of the Media

From the perspective of the public domain of the news media, the ideal public domain is by no means a separate and general public concept, but can be open to the disadvantaged to express different opinions, respecting the different expressions, and enriching the diversity of public discourse (Jinhua, 1995). But this ideal public domain is difficult to appear in the actual media, and the number of public opinions that the media express has a certain space or time limit. The main approaches for the public to express their opinions directly through newspapers are reader contributions, hosting layouts, letters from readers, public events, classified advertisements, etc.

The development of emerging communication tools, especially the development of the Internet, has enriched the political participation of the media, breaking through the bottleneck of traditional mass media such as newspapers, radio, television and magazines, which can only accommodate the comments and communication from a very small number of audience, and provide a space for most people to express their opinions and discuss. In this way the political participation of the media is enhanced. In recent years, the Chinese National Congress (including local conferences), Chinese citizens have actively participated in "online discussions", and many opinions and suggestions have flown to the National People's Congress and the CPPCC through emails like snowflakes. Before some major regulations and measures were formally promulgated, the government attached great importance to soliciting opinions through the Internet in advance and implemented an "online hearing". For example, in 2000, the State Planning Commission publicly solicited opinions of people across the country on the draft National Tenth Five-year Plan, which attracted active participation from all over the country. Here, one of the prerequisites for people's active participation and expression of opinion is to believe that this channel works.

In short, credibility is a bridge for realizing political participation, which affects the dissemination of political information and the people's behavior of expressing opinions in the media. The role of media credibility in political participation is analyzed in a report on *Zaobao* of Singapore.

> On March 1st, the "Zaobao Communication" edition published a contribution entitled "The Reader's Words Work". The reader enumerated several suggestions given through Zao bao,

all of which were quickly responded by the relevant government departments, and the problems were solved in time.

He personally realized that "this is a silhouette of Singapore's democratic society. Newspapers support the people to speak. People's opinions in the newspapers are valued by the relevant authorities, so that the problems can be resolved in a timely manner and have a profound impact ...

On the surface, this contribution merely expresses good communication between the relevant government authorities and the people. However, further analysis finds that media credibility actually acts as the bridge behind this benign communication.

First of all, people must believe that the media are willing to plead for the people, and it will expose the suffering of people. Then they will inform some problems and situations around them to the media. It is hoped that the media's "muffler" will attract the attention of the relevant authorities so that the problems can be solved. This is the embodiment of media credibility on the people.

Correspondingly, as the government has a good understanding of the influence of the media among the people, it will pay attention to relevant reports, opinions and speeches in the media. Once problems related to itself have arisen, the relevant ministries will further verify and manage to quickly solve them. This is a manifestation of the government's trust in the authority of the media and the authenticity of media readers' statements (Shu, 2003).

5.2.4 The Impact of Credibility on Public Opinion Supervision

5.2.4.1 Public Opinion Supervision on the Media

Journalist Feng (1999: 7) believes that the public opinion supervision is the embodiment of the right to freedom of speech and freedom of the press guaranteed by the Constitution and it mainly supervises the exercise of power by public authorities or organizations through the media or other news sources. It is a political right concerned with the participation of individuals in public affairs. Ai Feng positions the public opinion supervision in China from the level of the entire power supervision system. He argues that with the advance of China's socialist democracy, the power supervision systems are gradually established and improved. Intra-party supervision, people's congress supervision, administrative supervision, democratic supervision and public opinion supervision have constituted the six pillars of power supervision in China. And public opinion supervision is the most special type of supervision in the six major supervision systems. Although it is not compulsory and binding, it is the most powerful supervision. This supervision is valued by modern democratic countries in the world. It is the most direct public supervision.

Regarding the role of mass media in the public opinion supervision, Feng (1999: 5–7) thinks that the public opinion supervision, as the exercise of supervision of political power on behalf of public opinion, is one of the important political functions of mass media. First, mass media has promoted "transparency in the political process" through public reporting on the real life. Increased political transparency has placed government workers, officials and all other people in power under the eyes of the public, thereby increasing the moral pressure on corruption and abuse of power. Second, mass media supervise corruption, breach of duty, abuse of power and other misconducts in the exercise of public power, which can serve as a "social exhaust valve" and contribute to the political stability of our society.

In other words, the media's public opinion supervision is shown in acting as a channel between the subject of supervision (the people) and the object of supervision (public power). It is a representation of the public opinion. In the power supervision system of our entire society, the public opinion supervision in the news media is a kind of supervision that can reflect public opinion and has a wide range of influences. The extent to which such supervision is truly implemented is directly related to the construction of modern democracy in our society and the maintenance of social stability.

5.2.4.2 Three Types of Public Opinion Supervision Reflecting Media Credibility

Parents and teachers often tell children the "wolf is coming" story. It says, a child loved pranks and often shouted, "Help! Help! A wolf is coming". The adult villagers rushed to his help. They found no wolf there. The boy laughed at them. He said that he had cried for the sake of fun. They went back. On another day, he repeated the joke. The villagers again came to help him. No wolf was found. One day a wolf came there. When the boy shouted, "A wolf is coming", but no one trusted him and came to his help. The wolf killed him.

If the media is likened to the child above, the public would be the adult. If the media cannot trust the public, its public opinion supervision will lose its foundation for survival. In other words, the credibility of the media itself is a key factor that determines the degree of implementation of the media's public opinion supervision. The influence of media credibility on the effect of public opinion supervision is manifested in the subject and object of supervision. If media credibility is low, the media will be unreliable for the public, and their enthusiasm for participating in the supervision of public opinion will be reduced, so that the number of people engaged in public opinion supervision will become smaller. If the public does not have high trust in media opinion supervision and does not respond positively to it, the effectiveness of media opinion supervision will be weakened. For the public power, the supervised objects have more room for excuses since media credibility is low. For the entire power system there is less pressure to feel supervised, and the degree of concern for supervision becomes weaker. As a result, public opinion supervision will lose its due role.

Only credible media can produce the effect of public opinion supervision in three types.

(1) The media have higher credibility because it is considered by the public to have a higher level of professionalism, which can play a great role in the public opinion supervision. This type of media has long demonstrated the courage

to face real problems and maintained an objective and fair image, leaving the public with the impression of pursuing the ideal of journalism. They build their own credibility through the public's recognition of their professionalism, and become influential in society.

The media has such credibility that its supervision has a wide range of influences among the people, and the target of supervision is under great pressure. Such media are developed abroad for years, including *The New York Times*, *The Wall Street Journal*, *The Times*, *Los Angeles Times*. In the past two years, although *The New York Times* committed frequent acts of journalistic fraud such as reporter Blair and damaged its credibility. But from another perspective, *The New York Times* has the courage to expose its own weakness, which is the best choice to maintain its credibility. It is precisely because of this professionalism of maintaining credibility that *The New York Times* maintains its status as a century-old mainstream newspaper. These newspapers with high credibility were able to exert their influence in the public opinion supervision, such as the Nixon Incident in the United States and the Clinton Scandal, and became "great works" in the supervision of American media.

Southern Weekly in China can be regarded as a newspaper that has established high credibility among the people through its professional pursuit. Its critical supervision has a wide influence in the country. It has an impact on society. The government attaches importance to its reports.

(2) Media credibility is mainly established through the executive power it represents. Compared with the last category of the media, this kind of media supervision can receive government attention in a relatively short time, and the problems can be solved to a large extent. The people trust it because of its authority. *Focus Interview* program is a good example in China. From the broadcast starting on April 1, 1994, the program had been almost "unbeatable." It broadcast at 19:38 every day for 11 min and 50 s, and had high ratings. There are more than 60 programs similar to *Focus Interview* in China, but none of them can have such influence as *Focus Interview*. It is partly due to the quality of the program. Besides, it is broadcast by the highest-level TV stations in the country and the issues it reported can receive more attention from the central and local governments.

For this phenomenon, Linhao (2004) argues, CCTV's reports have such a good effect, which is inseparable from its unique status in the country. The popularity of CCTV's public opinion supervision program is a reflection of the effect of official rank standard. Strictly speaking, although CCTV is a national television station, their staff are grouped into the national civil servants. It seems that it cannot be called the official rank standard, but the power of the hierarchy is reflected in the press. In China, the media are classified according to the hierarchical levels of county, city, province, and central government. CCTV is undoubtedly the highest media in the television field. Its status makes it "can say what others can't say" and the influence of its programs is much greater.

Moreover, Mo Linhao believes that due to the influence of local protectionism and other factors, the local media have many taboos on some negative local reports.

CCTV cannot be subject to such taboos, because local governments have no power to restrict the central media. Even when covering the same news subject, the influence of CCTV shows is significantly greater than that of local media. For example, the incident of inferior milk powder in Fuyang was reported in the local media for a long time, but it was not until the reporting of CCTV that the news aroused widespread national attention, which helped to solve the problem. It can be seen that there is the effect of official rank standard on it.

(3) The Internet, especially the Internet forum, has some characteristics of the public domain: openness, equality, legality, free access, expression of opinions, and the field of public opinion. Internet users can easily express their opinions, and can get some information that the traditional mass media cannot control. The Internet is a communication channel that seems to be controlled by gatekeepers without professional training. It is generally thought that its credibility will not be high among netizens. On the contrary, this kind of communication channel, with the connotation of the public domain, has high credibility among netizens.

An online survey conducted by Zero Survey from May to July 2003 found that the Internet became the main channel for the Internet users to obtain information. Its credibility was second only to television and it was the second largest media trusted by Internet users. Among all kinds of online information, news enjoys the highest trust. From the perspective of Internet users' trust in various media, the status of the Internet is not low. The netizens who believed that the network information was the most reliable accounted for 24.8% of the total. Although it was lower than TV (49.5%), it was significantly higher than the traditional media such as newspapers (17.6%) and radio (2.6%). Researchers think that the Internet enjoys high credibility on the whole, which shows that the impact of fake news, entertainment, gossip and other misinformation on netizens that are often present is not conclusive. It is worth noting that the credibility of the Internet is much lower than that of TV, and the advantages of the Internet should be based on the establishment of appropriate monitoring mechanisms to further enhance the authenticity in order to be more credible (Shuxiang, 2003).

It is because of high credibility of the Internet that the voices from the Internet can form a strong public opinion supervision. In the past few years, China's Internet has become the most popular media supervision, which surpass television and newspapers. The BMW incidents in Harbin and Changsha, the Sun Zhigang incident, all formed a huge public opinion field on the Internet, which finally attracted great attention from officials.

5.2.5 *The Impact of Credibility on the Political Communications*

5.2.5.1 Political Communication of the Media

Simply put, political communication refers to the transmission of political information in the political system. According to the American Scholars Almond and Powell, political communication can be divided into five categories. The first is informal or primary structures. The second is non-governmental structures, such as family, economic or religious institutions. The third is the input structures, such as political parties and other professional organizations. The fourth is the output structures such as the chief executive, the legislature and the executive branch of the government. The Fifth is the means of mass communication such as television, radio, newspapers, and magazines (Almond et al., 1987: 168).

Among the five categories of communication, mass media is the most popular and timely type of communication. It can spread the information to the widest range in the shortest time at a minimum price. Therefore, in the study of contemporary political communication theory in the West, the model of research focusing on traditional communication media has been established.

The political communication of mass media plays an important role in the political operation of modern democratic countries, the communication of peer political groups and the direction of policy between countries. For political parties within a country, mass media are often used as a bridge for exchanging opinions. Similarly, the communication function of mass media between countries is reflected in different levels.

5.2.5.2 The Role of Media Credibility in Reducing the Cost of Political Communication

American scholar McPhail expressed his doubts and anxieties about realistic political communication when studying the issue of racial discrimination and racial conflict in the United States. McPhail believes that the general form of political communication is oppositional or antagonistic, which is characterized by affirming itself and negating the other. A very likely consequence of the oppositional political communication is away from the communicating parties. However, the cooperation and integration of opposing communication parties may happen. The premise for this result is that in the process of oppositional political communication a political dialogue occurs, which can lead to political cooperation and integration (Weimin, 2001).

The results of these two kinds of communication are demonstrated in the model, as shown in Fig. 5.3.

For the above two forms of communication, it is obvious that the first "oppositional communication" to "political divide" is the worst option for communication. The

Oppositional communication → Political divide
Oppositional communication → Political dialogue → Political integration

Fig. 5.3 The results of communication

second model of "oppositional communication" to "political dialogue" to "political integration" is the best result, in which mass media play the role of political dialogue.

The study of communication has developed from the earliest media effect to today's audience effect, that is, it is not any media report that will achieve the corresponding effect, but it depends on the audience's acceptance. Therefore, the premise for the government policies and decisions to be understood and supported by the people through mass media is that the media is credible.

If media credibility is not high and the public's perception of the media is skeptical, the various policy information displayed by the government through media reports will hardly attract public interest. As for the attitude of the public, for some contradictions in the society, the public opinions on the government, or policies and behaviors that are opposed by the public are reflected in the government's persuasive reports of the media. If the credibility of the media is not high, it is impossible for them to achieve the effect of changing attitudes. It's like in interpersonal communication, when an untrustworthy person and a trusted person send the same information, the receiver will respond differently. For the untrustworthy people, the receiver may have to doubt whether it is true or not, and will verify it under suspicion. As for the information given by trusted people, the degree of suspicion of authenticity is very low. For persuasion of the untrustworthy people and the trusted ones, the results are similar to that of sending information. Persuasion of untrustworthy people may not have any effect at all, or cause an opposite effect.

So the government needs to take more time and cost in communicating with the people. If media credibility is high, these extra costs will be reduced. And another problem is that if media credibility is not high, the communication is not effective, and the conflicts are not resolved in a timely manner, which may exacerbate the conflict of opinions. It will undoubtedly increase the new communication costs.

In China, where the economy and society are currently in transition, many new problems arise. The new government policies and regulations need to be understood by the public. There will be some contradictions. The political communication of mass media is extremely important. The model of "oppositional communication" to "political dialogue" to "political integration" is needed to make the public understand the new policies and support these policies. It has an impact on the whole reform process. For example, in the state-owned enterprise reform that began in the mid-1980s, mass media reported the government policies to thousands of laid-off workers, making them realize the significance of such reforms. In this case the media is required to be credible.

5.2.6 *The Impact of Credibility on the Political Control of the Media*

5.2.6.1 Political Control of the Media

From the perspective of system control, political control is a means to maintain national order. It is an organic part of social control, like moral control and educational control, requiring the social constituents (individual, group, organization, etc.) to follow social practice and norms.

Regarding social control, Marx argues: The ideas of the ruling class are in every epoch the ruling ideas, i.e. the class which is the ruling material force of society, is at the same time its ruling intellectual force. The class which has the means of material production at its disposal, has control at the same time over the means of mental production. Generally speaking, the ideas of those who lack the means of mental production are subject to it (Yunhu & Yunlong, 1989: 33). As for the relationship between the media and political control, in Marx's view, control of productive resources and communication media is a source of social power in capitalist society. It is this control that enables one class to exploit other classes without such control and to maintain economic, social and political inequality. It is this control that makes a class become the ruling class (Yunhu & Yunlong, 1989: 20). That is to say, out of control of media opinion in a country means out of control of the regime. Therefore, no matter in socialist countries or capitalist countries, mass media are an important means of political control. The political control of mass media directly affects the stability of the political system.

5.2.6.2 The Decisive Effect of Credibility on the Political Control of the Media

Only when the public trusts the media can the political control of the media be effective. The *BBC Handbook 1941* explicitly stresses that the long-term propaganda policy of the modern world cannot be based on lies (Carruthers, 2002: 94). In the Second World War the British news media sought to maintain the trust of the public during the war, maintaining the reputation of honesty, freedom and authenticity among British people abroad under the premise of filtering confidential information. This made the British propaganda campaign work well, and played a big role in the victory of World War II on another front. During the war, in Europe including Germany, the BBC radio was popular despite many penalties for listening. In the UK, even right-wing commentators such as George Orwell, who disliked the BBC's sluggish style, acknowledged its "authenticity" and commented that the BBC was more credible than newspapers. S. Nicholas remarked, "when people think that the British are just being told what the leaders want them to know, and there are many deliberately mistaken or even refused to tell them what they really want to know", the BBC's praise is strange (Carruthers, 2002: 118–119).

For China, the "mouthpiece" attribute of mass media itself determines the importance of political control. Media credibility determines the role of political control. Let's take a look at two cases during the SARS outbreak to see how credibility determines the media's political control.

> When the atypical pneumonia virus broke out suddenly,.. "I feel that the actual situation is far more serious than the official report, and suspect it is concealed. In addition, I have doubts about the government's means, capabilities and measures to contain the spread of the disease. This worry put me in a state of inexplicable anxiety." A netizen named "nkfeiyang" posted a message online to ask a psychologist for help (Legal Daily, 2003).

Another case is the SARS incident in Guangzhou in February 2003 (Hongzhong, 2004: 12).

In early and mid-January 2003 atypical pneumonia cases appeared in Heyuan and Zhongshan of Guangdong Province, and there were rumors of "flu deaths". Before the Spring Festival, atypical pneumonia cases also occurred in Guangzhou and infected some medical staff. On February 8th after the Spring Festival when was the first working day after the national holiday, the incident became "fatal flu occurring in Guangzhou through mobile phone text messages and word of mouth. This information was quickly spread among Guangzhou residents. Immediately, many Cantonese started looking for this information online. In terms of the number of deaths and the cause of illness, posts with different opinions appeared on some online forums. The following two days the word spread rapidly and became more and more sensational. It finally erupted into a panic across the city and caused the vast majority of families to be swept up by a large-scale rush to buy Banlangen (traditional Chinese medicine), white vinegar and masks.

On the morning of February 10, police officers in some public places such as the railway station began to wear olive-green masks.

On February 11, the Guangzhou Municipal Government held a press conference on this incident. In the afternoon, the Guangdong Provincial Department of Health also held a press conference. On the 11th and 12th, major newspapers in Guangzhou started to report it. The large-scale reports of the media quickly curbed the panic that had spread among the citizens for many days, and the snap-up of Banlangen, white vinegar and masks subsided.

Guangzhou residents, who had just experienced a crisis, were shocked. Suddenly, on the afternoon of the 12th, they were involved in a large-scale snapping of rice and salt. On the 13th, Guangzhou Daily made a headline on the front page "Shortage of salt and rice is nonsense", with a large space to report the wave of rice salt buying in Guangzhou and other cities. As soon as the news came out, the new wave of snap-ups subsided that afternoon. A summary of the communication model in the process is shown in Fig. 5.4.

It can be seen from the model in Fig. 5.4 that throughout the incident people have multiple channels for obtaining public information, including official mass communication channels and diverse folk media. Our mass media is only part of the many channels. Our mass communication is in an environment that competes with other communication channels.

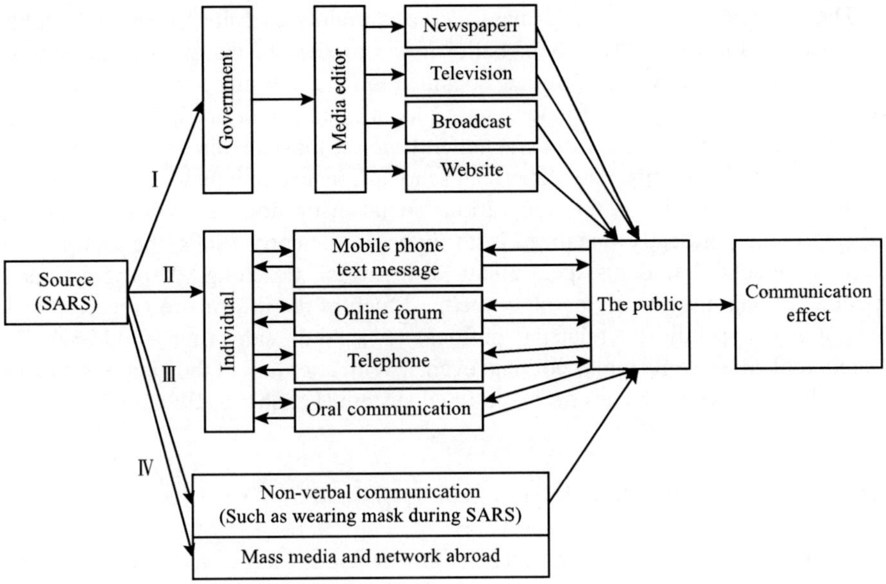

Fig. 5.4 The communication model

The above events are typical manifestations of the media control function. In the incident, as long as mass media follows the basic facts to report, high credibility of our mass media can give full play to its political control. Why can mass media play a big role in control when facing the reality, rumors and gossip that are related to personal safety? The answer is the public trust. Therefore, media credibility is a decisive factor in determining whether its political control works.

5.2.7 *Influence of Credibility on the Socialization Function of Media*

5.2.7.1 Political Socialization of the Media

Political socialization is the process by which people form their ideas about political orientations and behaviors through learning and practice. Socialization processes in different political systems lead to different political orientations and make people have different political experiences. Some scholars have summarized it into five aspects: (1) strengthening the individual political behavior and model; (2) changing the individual political behavior and model; (3) communicate political culture; (4) developing the "empathy" of the people; (5) positive and negative influence on political participation (Lansheng, 1996: 408–412).

The political socialization of mass media is mainly to cultivate people's political ideas and shape political personality in the process of disseminating political information. In other words, political socialization is a learning process. In real life, besides school, family and friends, mass media is an important learning channel. "For a teenager, he or she spends as much time on mass media as he or she does on school and family life. The American scholar Greenstein believes that if schools communicate information through various social institutions, so will mass media, especially the politically socialized institutions. In the early 1960s, the average US elementary school students spent about 30 h a week watching TV programs and spent additional time receiving other media. Much of the information received was not politically specific, but related to political news. At the same time, children deliberately looked for some major political events. With attention of the media, they will gradually perceive the basics of the political system (Lansheng, 1996: 407).

5.2.7.2 Credibility Affects the Political Socialization of the Media

Although children do not have much judgment about media credibility, they can make indirect choices through the influence of the people around them, especially their parents. If parents feel that the media is not credible, their judgment will affect their children's exposure to the media and their choice of media. And parents' media contacts have an indirect effect on children's modernity. It was confirmed by Chinese scholar Bu Wei's research on Zhejiang and Beijing. She found that parents' media exposure and education level are major factors affecting children's media exposure. The higher the level of parental media exposure, the more often children are exposed to books, newspapers and magazines. To preference for media content, parental media exposure is still an important factor. Parental exposure to media content affects children's (Wei, 2002: 290–303).

Media exposure and content choice will undoubtedly have an impact on children's future social norms and political ideas. The diverse communication channels have brought down the various political ideologies entrained. For those media that represent the mainstream political values of society, they are facing competition with various channels. Whether they will win the trust of people, and more specifically of parents, is crucial to shaping a social person that meets requirements of the future society and has a sense of modern democracy.

There is the issue of political socialization for adults. Harvard's well-known scholar Alex Inkeles (Wei, 2002: 263) found in the study of "the modernization of man" that substantial changes can be made in a man's personality or character at any stage of life, especially in the early and mid-stages of adult with appropriate environmental stimulation and social support, rather than what is commonly believed by psychologists that all important things in personality development occur before the age of 6 and are stereotyped before the age of 16.

5.2.8 *Influence of Credibility on Government Image of the Media*

This function is mainly discussed in the context of Chinese media. The "mouthpiece" of the media in China was discussed in the previous chapters. In the eyes of the public the media and the government are actually bound together. The media represents the Party and the government, and its performance is directly related to the image of the Party and the government.

Media's functions of the government's image are generally classified into four categories according to the degree of marketization of the media. The first is the media such as the Party newspapers, the Party journals, China Central Television and China National Radio which have the strongest function of the government image. For example, *People's Daily* represents the voice of the Party Central Committee, and its front page editorial indicates a policy guide for the public. The second is a comprehensive or current political medium with a relatively high degree of marketization. In terms of the government image this type of media is not as strong as the Party newspapers and journals. Its entertainment and special issues are more popular, but such media has a large amount of communication. It acts as an important bridge between the government and the people. The role of the government image played by the media has a greater influence. The third is some industry media, which are generally sponsored by relevant industry agencies and present the industry policies and events. Although its influence of the government image is not as broad as that of the second media, it shows the agency image. The fourth is market-oriented media of entertainment, sports and health, which relatively have the weak function of government image.

The influence of the credibility of the above four types of media on the government image is distinguished in order from strong to weak. For Party newspapers and journals, the decline in credibility has the largest impact on the government image, while the media such as entertainment, sports, and health have the smallest impact.

In general, due to the "mouthpiece" attribute of China's media, regardless of the type of media, the decrease of credibility will damage the image of the government. The "bundling" of the Chinese media with the government is a double-edged sword. The media can rely on the government authority to enhance credibility and make better use of the political functions of the media, just like the current credibility of the media in China. The decline of media credibility will undoubtedly affect the authority of the government.

5.3 Rationality—Premise of Positive Impact of Credibility on Political Functions of the Media

The influences of credibility on the political functions of the media have been discussed above. One general idea is that the reduction of credibility will have a negative effect on the political functions of the media. For today's China, due to its high credibility, the political functions of the media can be brought into play. On the one hand, we should pay close attention to the danger of reduced credibility. On the other hand, we need to pay attention to the harm caused by blind trust in the media. This book specifically addresses the issue of rational consciousness of media credibility.

Chinese scholar Lidan (1999: 22–23) points out the quality of public opinion. He believes that public opinion is a natural form of group opinion, and therefore has strong spontaneity and blindness. It is passive to a certain extent, that is, public opinion contains both rational and non-rational ideas. Similarly, media credibility has an issue of quality, covering elements of rational and irrational consciousness. Regarding media credibility, we cannot simply look at its level (high or low). We must examine the size of the rational consciousness in media credibility. Only media credibility that contains more rational consciousness is of high quality. Only in this way can high-quality credibility exert a positive influence on the political function of the media.

Reviewing the performance of the media during the Great Leap Forward and the "Cultural Revolution" in China, or the news reports were obviously not in line with the facts, although there is no consistency in the reports. Sometimes the reports on the same thing or the same person are quite the opposite, or not in line with common sense. But the words used by the media are still one of the main channels driving the national bursts of movement. The people seem to be crazy for trusting the media.

This phenomenon is caused by the historical, political, social and individual factors. As far as the credibility of the media itself, we need to think whether this kind of blind worship of media is conducive to the political function of media. Obviously, media credibility under the premise of rational and critical consciousness is conducive to the active exertion of the political function, and under the premise of blind trust it is not conducive to the healthy development of the media's political function. The higher the credibility of a medium containing more irrational consciousness, the greater the negative effect it may have on its political function.

The trust in the media needs to have a rational and critical consciousness, which has very practical significance for the political function of the media in China. In addition to being of great significance to the construction of modern political civilization, it plays the following specific roles. First, the rapid development of the Internet has changed the form of gatekeepers in traditional media, and various small news and rumors appear on the internet. In order to keep these gossips and rumors from causing confusion in public perception, the rational criticism of netizens who have access to them is very important. If the netizens' trust in the network is rational, they can identify them, and these gossips will not cause much harm. However, if netizens' trust in the internet is blind, various gossips on the internet will have an adverse effect. Second, in the process of marketization, in order to obtain more economic benefits, some media with a high degree of marketization will use kitsch reports, or excessively report some self-interested and materialistic contents. It is essential for the audience to have a higher rational consciousness. The public rational trust in the media can make the mainstream values of the media take effect. Third, with the increase of communication, some overseas media or illegal publications will flow into the country. How to identify right and wrong and not to be misled? The first thing we should do is not to blindly trust, but to have a sense of rational judgment.

5.4 Impact of Credibility on Economic Functions of the Media

Schramm believes that the economic function of the media is to provide information on resources and opportunities to buy and sell, to explain information, to formulate economic policies, to activate and manage markets, and to create economic activities. Among these functions, explaining information and formulating economic policies were discussed in the previous section. Here, the effects of credibility on the economy are discussed in two aspects. The first is the effect in the entire economic system, that is, the role in economic transactions, which is what Schramm said about resources information and buying-and-selling opportunity. The second is the effect on the development of the media market, that is, the role in the development of the media industry, which is to create economic activities according to Schramm.

To discuss the economic role of media credibility, we first need to examine the role of trust in the economy. There are many expositions on the economic role of trust in sociology and economics. This book summarizes them into three aspects, i.e. the role of lubricants in trade, the simplification of decision-making and the constraint of transactions. The three economic effects of trust are discussed separately below. Then the influences of credibility on two aspects of economic function are analyzed.

5.4.1 The Role of Trust in Economy

5.4.1.1 The Role of Lubricants

In the 1970s, the role of trust in the economy gradually gained attention, and the concept of trust began to be accepted by mainstream economists. A more general view among economists is that trust is the lubricant in transactions. Kenneth Arrow, an American economist and the winner of the Nobel Prize in Economics, thinks that nothing is more valuable than trust. Trust is an important lubricant of a social system. It is extremely efficient, since it saves a lot of trouble to have a fair degree of reliance on other people's word. Unfortunately, this is not a commodity which can be bought very easily. If you have to buy it, it means that you have some doubts about what you've bought. Trust and similar values, loyalty or truth-telling are examples of what the economist would call "externalities". They are commodities, which have real, practical, economic value. They increase the efficiency of the system, enabling you to produce more goods or more of whatever values you hold in high esteem.

Simply put, this lubricant effect of trust can reduce transaction costs. Some scholars have made some discussions on it, and find that those who have a strong form of trust relationship with other companies will enjoy a certain competitive advantage, and they can reduce the various costs associated with transactions between companies.

This lubricant's role of trust has found many practical applications in specific commercial activities. Brenkert proposes the concept of "marketing trust." He argues that trust promotes the relationships that people seek. Therefore, marketers try to develop trust relationships with their different customers. For example, Sears encourages potential customers in its financial services department to "trust Sears to work for you." The American Automobile Association urges its customers to "walk with people who trust you." Even foreign companies have joined the action, for example, India's Maruti Motors has announced that it will "build trust worldwide".

Regarding the role of trust in economic transactions, some sociologists believe that it is not just a lubricant, but a central role in the economy. As Filangieri pointed out, confidence is "the soul of commerce", and the credit alone can generate must be regarded as a "second species of money" (Pagden, 2000: 127–141). Trust is central to all transactions and yet economists rarely discuss the notion. It is treated rather as background environment, present whenever called upon, a sort of ever-ready lubricant that permits voluntary participation in production and exchange (Dasgupta, 2000: 49–72).

In short, trust plays a very important role in economic transactions and is an indispensable part of reducing transaction costs in economic activities.

5.4.1.2 The Role of the Simple Mechanism

German sociologist Miklas Lumann puts forward in his book *Trust and Power* published in 1979 that trust is one of the complex mechanisms of social simplification. Luhmann thinks that the natural and social environment of human life is becoming increasingly complex, and a simple system must be established to deal with them, such as language, currency, scores, and test systems. Relying on simple mechanisms is a strategy for human survival, and trust is a simple system. So Luhmann states, by reduction of complexity, trust makes it possible to interact (Hai, 2004).

The simple mechanism of trust is that trust can reduce the time and risk of trading decisions, greatly reducing the interference factors outside the transaction itself, and making the decision simple and fast. As some have argued, in the face of the problem of time difference, trust cannot reduce time to make the promises immediately fulfilled, but the trust relationship has an effect of time reduction. The same is true of the risk of uncertainty. People must face uncertainty normally and make decisions and take action without uncertainty. Tossing a coin is a simple means of determination, and trust is also a simple mechanism to help make decisions. To trust someone is to discard the uncertainty from a subjective perspective, and to leave those uncertain factors to the trusted person, believing that he will react appropriately or when uncertainty arises, the trust relationship between two parties will definitely lead to a good solution. Economists like to examine the relationship of trust from the benefits of reducing transaction costs. This is actually the effect of simplification. Trust greatly reduces the cost of dealing with time risks, moral risks and uncertainty risks (Hai, 2004).

5.4.1.3 The Role of Constraint

Some scholars believe that there is a contract, and trust does not play a role in transactions, because contracts can restrict the performance of transactions. Haas and Deseran argues that trust is not necessary to business transaction because a legally binding contract replaces the need for trust between buyers and sellers.

In fact, trust is actually a constraint mechanism. Because of the trust relationship, the transaction is not a simple relationship of buying and selling, but there are many factors other than pure buying and selling, which have a restrictive effect on transactions. The written contract is only a manifestation of the trust mechanism and an external form of the constraint of trust. Bluhm thinks that trust is a prerequisite for business transactions, especially when there is more than one transaction between a buyer and a seller. Bloom argues that since distrust is a problem in complex societies, a written contract is nothing more than a manifestation of a so-called trust mechanism—a common practice that shows individuals interacting according to a certain understanding. A written contract guarantees, not replaces trust.

Furthermore, some scholars believe that trust is one of the most effective binding forces in buyer/seller relationships. The detailed text of the formal contract, the thick legal materials prepared for the claim, and the various forms of recourse will not give the buyer pure and basic trust between the salesman and the group he or she represents. Establishing trading relationships has such high expectations. Some Chinese sayings in shopping malls reflect the role of trust, such as "business is off while friendship is on", "make friends first, do business later".

5.4.1.4 The Role of Trust in the Economy

Trust plays a role of lubricant, simple mechanism and constraint in transactions. How do these functions work in economic activities? It can be seen from the research conducted by some economists.

Economist Zhang Weiying conducted an analysis of economic data related to trust and economic indicators including regional GDP, business scale, business efficiency, product sales profit, total profit, development speed of private enterprise, average size, foreign investment, and found there was a correlation between trust and these indicators. This was a data analysis based on a survey of entrepreneurs across the country in 2000.

This data analysis found that there was a positive correlation between per capita GDP and the trust in all provinces in China, that is, the group with the highest per capita income has higher trust, and the group with the lowest per capita GDP has the lowest average trust. The level of trust between the two is ten times different. There is a strong correlation between per capita GDP growth and trust.

Trust also has a relationship with business size. The correlation coefficient between trust and the output value of state-owned enterprises and non-state-owned enterprises above designated size is 0.55, and the correlation coefficient between the output value of state-owned and state-controlled enterprises is 0.63.

Trust not only affects the size and distribution of an enterprise, but also affects its effectiveness. Trust is closely related to the profit of product sales of state-owned and state-controlled enterprises (0.59). The correlation with the total labor productivity of state-owned enterprises is significant (0.67), and the correlation with product sales profits and total profit is significant (0.67 and 0.62).

The relationship between trust and the development speed and average size of private enterprises is obvious. In regions with high trust, there are more private enterprises per million people (the correlation is 0.89). There is the correlation between trust and the number of private enterprises, the number of employees, the number of investors and the number of private enterprises. The correlation coefficients are between 0.51 and 0.63. If the per capita GDP trust is standardized, the correlation coefficient between trust and the number of private enterprises is 0.80.

Trust has a significant impact on the introduction of foreign capital. This is because, as an organization that maximizes profits, foreign investors only favor areas with high trust. The correlation between trust and the number of foreign-funded enterprises, the total foreign investment, the registered capital of foreign-funded enterprises, the total foreign capital introduced per million populations is 0.50, 0.61, 0.63, 0.81 respectively.

5.4.2 Media Credibility as a Trust Agent in Economic Transactions

The role of trust as lubricants in economic transactions and simple mechanisms discussed above refers to a kind of interpersonal trust. As for media credibility, it has the same effect. However, the specific mechanism for media credibility to play a role in economic transactions is different from interpersonal trust. Interpersonal transactions occur from individuals to individuals. The two individuals transact because they trust each other. The transactions that occur through the media take place from individuals to media to individuals. The mechanism by which media credibility works is to act as an intermediary between the two individuals. A prerequisite for a transaction to occur is that the buyer trusts the media first and then trusts the media's commercial information or related business reports. In other words, media credibility plays the role of a trust agent.

At present the role of media credibility as a trust agent is increasingly significant, and most consumers learn about product information through media commercials. According to the "Survey on Consumer Acceptance of Commercials in China's

Cities" conducted by Zero Survey in July 2004 in ten major cities including Beijing and Shanghai, 56.2% of consumers admitted that they actively paid attention to commercials for learning about a new product in time, 51.5% of respondents "learned about relevant product information before buying" through commercials.

Besides the role of media credibility as a trust agent in economic transactions, it plays a role as a lubricant of trust and a simple mechanism. As for the restraint effect, since media credibility act as a trust agent, and it is not directly related to the transaction behavior, the restraint effect of media credibility is not significant. The following makes analysis of the role of media credibility as a lubricant and simple mechanism.

5.4.2.1 The Role of Lubricant

In the role of a trust agent, media credibility acts as a lubricant for transactions. High credibility of the media can reduce transaction costs. Because consumers can trust the products through their trust in the media, they can quickly and easily know the product, saving time and effort in understanding the information, and reducing the cost of confirming the information.

There are two aspects in consumer mistrust of commercials. First, because of their distrust in the media with low credibility, the consumers don't trust the media commercials. The second is trust in the media, but they do not trust the advertising content of the media. If media credibility is low, even if the commercials on the media can be known, their effectiveness will be greatly reduced. The audience still needs to reconfirm this media information from other channels, which increases costs. For commercials published by credible media, the audience's trust in them will be high. At this time media credibility plays the role of a lubricant in trading activities.

Here is an example. The TV sales direct originated in the United States in the 1980s. Since it is a convenient, low-cost and new sales approach, it quickly became popular in developed countries such as Europe, America and Japan. In the mid-to-late 1990s, a total revenue of the TV sales direct in the United States were around US$6 billion, accounting for 2% of the US retail industry, and in Japan it was close to US$6 billion. After being introduced to China in 1995, this sales approach developed rapidly at the beginning. But after three years, there arose a problem. According to "A survey of 1,230 consumers in six major cities including Beijing and Shanghai at the end of 1998 showed that more than 80% of consumers knew TV sales direct, only 4% of them were interested in it, and 71% of consumers were "no longer interested" in it or "not interested at all". The major reasons were that some commercials played to the gallery and were overpriced. In general, the key point is a crisis of trust (Xiong & Haoyuan, 1995).

There is one question that needs to be answered. Since media credibility takes the role of a lubricant for economic transactions, why are many popular newspapers not as credible as some serious newspapers, but they attract more commercials and charge higher prices?

For example, in China the provincial Party newspaper has higher credibility than its subsidiary newspaper, urban newspapers, but its advertising is far less than its subsidiary newspaper. This can be explained from three aspects. First, the functions of these two newspapers are different to readers. The Party newspaper is more upward and the urban newspaper is downward. That is, under the basic political framework, the Party newspaper is government-oriented and has a low degree of marketization. The target audiences of urban newspapers are ordinary people, and the degree of its marketization is relatively high. They are different types of newspapers in the market. In the same type of newspaper, the credibility is high, and the effect on commercials has a positive correlation. Second, the number of readers is a major factor in determining the amount of commercials. For example, in the same distribution area, out of two newspapers with similar circulation the one that has high credibility must have higher advertising prices and a larger number of commercials. Third, it is important that the quality of advertising content is highly correlated with credibility. For example, Commercials of some big brand need to be published in national newspapers with high credibility, such as *Southern Weekly*, to establish a good image. It is unlikely to be published in the same type of newspapers with low credibility no matter how cheap the price is. Similarly, in a city's real estate advertising, it is unlikely for high-end real estate advertising to be published on a medium with low credibility. The advertising on a medium with low credibility and a large circulation is targeted to relatively low-end brands.

5.4.2.2 The Role of Simple Mechanism

The role of media credibility as the simple mechanism for decision-making process is viewed from the perspective of sellers and buyers.

For the seller, the effect of media credibility on economic transactions depends not only on whether the buyer trusts the media, but also on the seller's trust in the media. If the seller does not trust the media, he or she will be alert to the media. This alert makes it impossible for sellers to take full advantage of the media's role of information transmission. In today's information society, if a product which cannot make full use of the media to convey information is promoted through other channels, it will take more time and money to achieve the effect of the media. For the seller, the high credibility of the media is undoubtedly a simple form of economic transactions. Sellers believe in the media, which means they actually believe in the consumers behind the media, so that the sellers will use the media to promote their products.

The following example illustrates that once the media fail to gain trust of the seller, the seller will be alert, and thereby reject the media.

> According to China Press Journal on Journal 5, 2002, Luoyang Copper Processing Group Co., Ltd. (Hereafter "Luo Copper"), the only company in China that provides coinage materials for the euro, refused numerous media promotions after winning the tender. In common sense, this is a good opportunity for enterprises to create momentum and make free advertisements. But Luo Copper clearly stipulated that this matter would no longer be publicized. Luo Copper said that instead of not propaganda, some boastful reports affected the integrity of

the company. Boasting made them ashamed in front of counterparts (Xin & Yang, 2003: 225).

From the above, it can be seen that if the media has high credibility, it is actually a matter of simplifying the process and reducing time cost of the promotion for the enterprise.

The same is true for buyers. If media credibility is not high, there will be no certainty about the commercial or business information reports on the media. Even if consumption is demanded, it will take more information to make a decision. If media credibility is high, consumers will not have doubts about the required information, which will simplify consumer decision-making process.

5.4.3 *"Scale" and "Valve"—the Influence of Credibility on the Audience of Different Media*

5.4.3.1 Media Credibility is Inconsistent with the Circulation of Newspapers

It is generally agreed that media credibility affects the media industry. Specifically, media credibility has effect on media consumption behavior. The higher media credibility is, the more audience there are. As Stephen Hess argues, journalism's interest in credibility is partly based on the common belief in "credibility sells newspaper". The subtext of this view is that media credibility is the basics of the survival of the media, and the media that loses credibility will lose the audience and the economic benefits.

According to the ninth issue of the *University Line*, a student journal of the Department of Journalism and Communication of the Chinese University of Hong Kong, published in November 1996, the department organized students to conduct a survey on the credibility of Hong Kong newspapers in September. The data showed that a large circulation did not mean high credibility, and there didn't seem to be a direct correlation between the circulation of newspapers and credibility. Is there any relationship between media credibility and its consumer behavior? What is the relationship between them? According to the survey data, the credibility rankings of Hong Kong newspapers are shown in Table 5.2:

5.4.3.2 Irrelevance Between Media Credibility and Content Preference

The above-mentioned problem of "inconsistency between media credibility and newspaper circulation" is the difference between the two concepts of content preference and credibility. The audience's preference for media content and their evaluation of media credibility are two issues. It is supported by empirical research of American scholars.

Table 5.2 Ranking of Hong Kong newspapers' credibility

Newspaper	Credibility index ranking	Ranking of the number of respondents who read the newspaper	SRH circulation ranking
Ming Pao	1	4	5
South China Morning Post	2	8	7
Sing Tao Daily	3	6	8
Standard	4		
Express	5		
HK Economic Journal	6	10	10
Oriental Daily News	7	1	1
Economic Daily	8	9	9
Sing Pao Daily News	9	3	3
Sing Tao Evening News	10		
HK Daily News	11	7	6
Apple Daily	12	2	2
HK Commercial Daily	13		
Tintin Daily News	14	5	4

The study conducted by Westly and Severin (1964: 325–335) in the early 1960s found the difference between media credibility and preference. It is clear that people 'did not necessarily assign greatest credibility to the medium they assigned greatest preference'. Trusting a medium and preferring a medium are different concepts, which is later proved by some studies. According to Stephen Hess, if there is a credibility problem in the journalism industry, it will be corrected (when it can be corrected), not because the problem of credibility will cause potential market loss, but because practitioners want to do their best work. A national survey of 1,002 adults by Tony Rimmer and David Weaver showed, "the frequency of one reading the newspaper and watching TV is not always related to his credibility evaluation of newspapers or TV news." "In fact, the Gannett Group report (Arlington, VA) 'no significant data' reflects changes in circulation that are based on efforts to increase the credibility of the media and its relationship with the members of the communities it serves".

The audience's evaluation of media content preferences and media credibility are two issues. There are some media content audiences may like, but their credibility is not high. For example, people just prefer some entertainment gossip news, and there is no requirement of credibility for these contents, just for fun. This can explain the irrelevance between the credibility of newspapers and sales volume found in the previous survey by the Chinese University of Hong Kong.

5.4.3.3 "Scale" and "Valve"

Since the audiences' evaluation of media credibility and media content preferences are two unrelated issues, in which aspects does media credibility play a role in the development of the media industry?

Media credibility does not have the same degree of impact on the market development of all media, but has different effects on media with different contents. If the media is simply divided into two categories: serious political affairs (represented as the pursuit of professionalism in the news) and entertainment (represented as the pursuit of the market, the professionalism of the news is not the core of media requirements), media credibility plays a different role in the reader market.

First, for a serious political medium, credibility is its market "scale", the credibility directly affects the audience ratings. Serious mainstream media such as the BBC in the United Kingdom and *The New York Times* in the United States. The occurrence of the incidents such as "Blair fraud" and "Kelly incident" affects the credibility of the media, which will hit the consumer confidence in these media, and may seriously affect the audience ratings. Some newspapers in China have made clear to establish their professionalism based on credibility and thus win audiences and popularity. For example, a newly-developed *The Bund* in Shanghai proposed a concept of news in the header: "Credit is life."

Second, for popular and entertainment media, media credibility is a "valve" for the market. The credibility that the public require for such media are relatively low. Their psychological expectations are fun for reading. It is the most important to satisfy the audiences' reading preferences. The most typical ones are *The Sun* in UK, some entertainment newspapers and entertainment TV and radio programs in Hong Kong, China. For example, some people are curious about the privacy of entertainment stars, or they like to treat their scandals as a "dessert". In order to cater to this strange hunting, some media will expose the so-called "magnificent material", whether it is true or not without verification, resulting in fake news. In addition, the increasingly fierce competition has made some media to offer the breaking news, to make up deliberately, to grab the attention of the audience and win the market. For such media, the credibility is not essential and they attach importance to the things that catches the eye.

However, low credibility that the public require for entertainment media does not mean that there is no requirement for credibility of these media, nor does it mean that credibility is not required. For such media, credibility is like a "valve". Within a certain range of credibility, it is not the main factor that determines its market, but once their credibility falls below a certain "value", it will have an effect on its market. For example, the biggest difference between entertainment newspapers sold in newsstands and some gossip newspapers sold on trains is that the news of the former is somewhat factual, and the contents on gossip newspapers may be 100% fake. This is a watershed. Once this boundary is crossed, newspapers are likely to be reduced to pure gossip ones and lose the basic requirement of satisfying readers curiosity, which will affect reader consumption of newspapers.

5.5 Impact of Credibility on Cultural Function of the Media

5.5.1 Inheritance and Shaping—the Cultural Function of the Media

Herder, a late 18th-century German thinker, proposes three basic characteristics of culture in his famous book *Outlines of a philosophy of the history of man.* First, culture is a pattern of social life, and its concept is homogeneous. Every word and every act, whether as a whole or in all aspects of social life, is an unquestionable part of "this" culture. Second, culture is a "national" culture. In the words of Herder culture is the "spirit" of a people. Third, culture has clear boundaries. As a regional culture, culture is clearly different from the culture of other regions. It can be said that these three characteristics have been regarded as the authoritative conclusions on cultural theory (Yang & Yi, 2000: 1).

But specifically, what is culture? This is a difficult issue to address. In *Culture: A Critical Review of Concepts and Definitions* (1963), the US anthropologists Alfred Kroeber and Clyde Kluckhohn enumerated over 150 definitions and proposed nine basic concepts of culture, i.e. philosophical, artistic, educational, psychological, historical, anthropological, sociological, ecological, and biological ones (Yang & Yi, 2000: 1–3).

According to the above definitions of culture, one of the three functions of media that Lasswell states in communication is the function of heritage inheritance, which is actually a cultural function. Just as some scholars have explained, the dissemination of cultural heritage from generation to generation is mainly to convey knowledge, values and social norms (Jinquan, 1996: 18).

In addition to inheriting culture, mass media is the shaper of culture. Like the rise of contemporary mass culture, the media plays a vital role in communication. There is the rapid development of network culture. A series of network cultural forms (such as blog culture) have begun to be popular with Internet users and spread to the society. As Stuart Hall points out in the article "Culture, Media and Ideological Effects", the primary cultural function of modern media is to choose to construct "social knowledge" and social images. The public perceive the world through this type of knowledge and images constructed by the media to appreciate the real life they have experienced. Therefore, in both the qualitative and quantitative aspects, in the twentieth century when capitalism was highly developed, the media established an important leadership in the cultural field. That is to say, as real life in modern society is becoming increasingly fragmented and irrelevant, mass media has become the main approach for atomized and fragmented social life to maintain a sense of "whole". Even in terms of economics and technology it has incomparable superiority over the older social communication tools in traditional culture. In contemporary life the huge figure of mass media is ubiquitous (Yang & Yi, 2000: 16–17).

Therefore, the cultural functions of mass media discussed in this book is presented in two aspects: inheritance and shaping.

5.5.2 The Impact of Credibility on the Inheritance of Media Culture

As the most extensive communication channel, mass media has a great influence on the inheritance of culture. One of the general functions of mass media summarized by Lassville is the function of inheritance. Like schools and families, mass media plays a role in transmitting knowledge, values and norms. The role of mass media's cultural heritage is mainly manifested in two aspects.

One is that mass media is a supplement to all types of education, and even directly and independently launches education. This is what Schramm believes is the role of teachers in mass media. Schramm (1990: 150) argues that for various forms of teaching, adult education and technical training, mass media can be extremely helpful. Where there is a shortage of teachers, trainers, and counsellors, the media can afford a considerable amount of teaching. And once the basic skills are learned, the media can provide opportunities for further learning. In the United Kingdom and the United States, this role is usually undertaken by the public television broadcast network, while in China, most radio and television stations at all levels provide various distance education programs, as well as online universities.

The second is that the daily reports of mass media are inheriting culture invisibly. The content of cultural heritage includes morality, values, codes of conduct, literature and art. For example, the social news of the China's urban daily refers to news about ordinary people. It is non-political, such as reports on fires, rescues, good or bad social phenomena. Among them, the reports on saving people, good or bad social phenomena are to convey a code of ethics, and readers can directly see from the reports what is good and what is unethical. Or these reports can stimulate public debate, trigger audience agendas of various sizes, and then pass on values through interpersonal discussions. In addition, the supplements of China's evening news, the broadcast of classics by radio stations and the opera columns on television are all inheriting culture.

There is no education without trust, which is a well-known truth in pedagogy. Regarding the cultural inheritance of mass media in various aspects of education, it is clear that media without credibility can hardly perform such functions.

Here it focuses on the impact of credibility on the second type of culture inheritance. The previous introduction of American communication scholars Vantaa and Hu used path analysis to find that credibility would lead to reliance, which in turn lead to exposure and to agenda-setting influence. Similarly, for the cultural heritage of the media, without media credibility the value and ethics reflected by the media news will have no effect of agenda-setting, and the traditional literature and art advocated by the media framework will not attract the audience's attention.

In addition, the media play a role in the process of socialization for children, which is an important aspect of the influence of media credibility on cultural heritage. The media here do not include completely popular media.

(1) Let's look at the exemplary role of the media for children. Chinese scholar Wei (2002: 389) believes that the media in the family play at least four roles,

which are the disseminator of social values, an important source of information and knowledge, the teacher of social learning, and the partner of entertainment. She further argues that the media are changing the old family education model. It changes the relationship between elders and juniors, for juniors may share information with elders at the same time. In other words, mass media plays an important role in the socialization of young people. It is an important approach outside school and family education, and an important channel for our social and cultural heritage. In a large number of media contacts, the role of demonstration helps adolescents understand many social values and behavioral norms.

Bu Wei made a further analysis of the media's role of demonstration. Social learning theory examines the impact of media on children from a long-term perspective, and believes that the human behavior is the result of learning. Learning behavior can be divided into two types: learning caused by the consequences of one's own behaviors and learning caused by the examples of others. Media influence falls into the latter category of learning activities. In social situations, most human behavior is learned by observation. From social observations, people form ideas about how new behaviors work, and this coded message becomes behavior-oriented. Television and other media are important sources of demonstration in modern society (Wei, 2002: 389).

(2) Let's look at the situation of children's exposure to the media in China. Here is a survey of the average weekly media exposure frequency of one child, shown in Table 5.3.

It can be seen that the media exposed to only children are extracurricular books and newspapers, followed by TV and audio tapes. Although 60% of children have a game console and 20% of children have a computer, the frequency of exposure is low. The researchers further find that the medium that only children spend most daily time on is extracurricular books, for 27 min, followed by TV for 26 min. The following in order are newspapers and audio tapes for 16 min, radio for 12 min,

Table 5.3 Survey data on the average weekly media exposure frequency of one child, Unit: %

Frequency	Newspaper	Extracurricular book	Magazine	TV	Radio	Audio tape	Game console	Computer
No	15.5	12.0	41.0	18.0	39.0	26.1	87.9	78.7
1 ~ 2 times	32.0	28.8	35.8	37.4	25.3	25.5	8.7	12.4
3 ~ 5 times	29.9	30.1	15.3	19.7	13.9	23.3	2.4	5.7
Daily	23.0	29.1	7.9	25.0	21.7	25.2	0.9	3.2
Number of Children (person)	3,276	3,284	3,261	3,268	3,273	3,267	3,276	3,247

Source Yunxiao & Wei (1998)

magazine for 11 min, Computer for 6 min, Game console for 3 min. Taken together, the only child is exposed to 3.6 media on average every day for a total of 118 min, nearly 2 h.

(3) With the development of economy and technology the influences of media on children's socialization are becoming greater and greater. In this context, whether the media can effectively function as a cultural heritage, their credibility is very important.

Children don't have much judgment on media credibility, and their media contact is influenced or even chosen by parents and teachers. Once media credibility is low, the perception of parents and teachers will affect children through precept and examples, and the media will lose its exemplary role. Such mistrust includes mistrust of media channels, or mistrust of sources, or mistrust of journalists and editors. Because parents and teachers do not have the time or expertise to evaluate the media as a whole, no matter which aspect of distrust or the point of distrust in the media will extend to the media evaluation.

This kind of mistrust in the media is prone to form stereotypes, which are difficult to change, and has a long-term impact on the role of media demonstration. If the public has a stereotype of trust in the media, that is, with high credibility the culture such as the values and traditional norms advocated by the media will provide a good example for the public and will achieve the cultural inheritance in the demonstration. If there is stereotype of distrust, that is, media credibility is low, it is difficult for the media to change this view of the public in the short term, affecting the socialization of children.

5.5.3 The Influence of Credibility on Shaping of Media Culture

5.5.3.1 The Shaping of Mass Culture

Some scholars have proposed the relationship between the media and contemporary culture from an aesthetic perspective, arguing that "media is the cradle of contemporary culture". "Images" have increasingly become a force that has a strong influence on public life and aesthetics, so that human cultural life begins with "reading culture" and return to a new form of "viewing culture" (Junjun, 2003).

The most popular discussion about the shaping of culture by the media is the shaping of mass culture. Not many scholars really regard media as culture shapers. They often take mass media as the main object of studying mass culture, and see what 'cultures' are carried by the media, and how these 'cultures' reflect the local social values, ideological trends, life and behavior at that time. In other words, they take the media as a mirror of society (Maozheng, 1994: 338). A major aspect of the media's role of shaping mass culture is its power of communication. Mass culture can take shape and widely spread, no doubt due to the power of mass communication (Maozheng, 1994: 336). The formation of mass culture is influenced by mass media. The 'mass culture' is the cultural content carried and conveyed by mass media (Shimin, 1983: 56–58)."

Since it was created, mass culture has developed with mass media. However, the popularity of mass media has been a controversial issue. The focus of discussion is on the characteristics of popularization in mass media products. Is it a good thing or a bad thing (Yang & Yi, 2000: 16–17)? There are two views on this issue.

(1) In the view of critical scholars of mass culture such as Max Horkheimer, mass culture is different from folk culture, and the Western theorists hold a critical attitude towards mass culture. Dwight Macdonald, an American social critic, argues that Folk Art grew from below. It is a spontaneous, autochthonous expression of the people, shaped by themselves to suit their own needs. Mass Culture is imposed from above. It is fabricated by technicians hired by business; its audiences are passive consumers, their participation limited to the choice between buying and not buying. Folk Art is the people's own institution, their private little garden walled off from the great formal park of their masters' High Culture. But Mass Culture breaks down the wall, integrating the masses into a debased form of High Culture and thus becoming an instrument of political domination. In the eyes of McDonald's, mass culture has become synonymous with standard culture, formula culture, repetitive culture, and superficial culture, and it has sacrificed many time-honored values for a false sense of happiness (Yang & Yi, 2000: 19–21).
(2) Scholars who have a positive view of mass culture believe that the cultural content carried and transmitted by mass media has the characteristics of being fast, large, popular, and direct, and it breaks the centralized monopoly of knowledge and culture. High Culture is not a patent for certain individuals. Popular newspapers, radio and television not only promote the advent of mass society, but also enable "mass culture" to flourish. Popular "folk art" draws nutrients from "high culture" and improves the quality of life of the public (Shimin, 1983: 56–58). In other words, although the mass culture shaped by mass media may erode our traditional culture, it enriches the quality of people's life and constructs our society. With the development of society, these roles of mass culture will be brought into full play.

One of the biggest problems with the critical theory of negative perception of mass culture is that it regards the public as a passively controlled object, completely ignoring the initiative of the audience in contact with the media. Judging from the empirical study of the communication effect, it still stays at the early stage of communication such as the "bullet theory". This point of critical theory of mass culture is obviously untenable. In this way the spontaneous demands of the people in mass culture are ignored. With the development of social democracy, more and more people from the lower class will enjoy the mass culture. However, it is necessary to pay attention to the issues of consumerism, superficiality, and erosion of traditional values.

5.5.3.2 The Influence of Credibility on the Construction of Media Society

Although there are some criticisms about mass culture shaped by the media, this shaping function of the media is present. In order to play its role of culture shaping, it is important for a medium to have high credibility. Regarding the specific role of culture shaping, Hall believes that the primary cultural function of modern media is to choose to construct "social knowledge" and social images. The masses use this type of knowledge and images to perceive the world and to appreciate what they have experienced. The following examines the culture shaping of media from the perspective of social construction of media.

(1) *Let's look at the influence of media construction on social cognition. Scholars have made in-depth discussions on this constructive role of the media by studying the relationship between mass media and people's cognition of violence. In particular, TV* shows have been concerned about the impact on the reality of the society because of their vividness and a strong sense of immediacy.

For example, since the 1950s, many researchers have studied the relationship between watching violent television programs and aggressive behavior of young adults and found that extensive viewing of television violence may cause children and adolescents to become more aggressive and anxious. Researchers explain that there are at least three reasons that can lead to aggressive behavior in children and adolescents. First, television violence shows viewers the world is full of violence. Some studies have shown that people who are exposed to more violent programs are more likely to feel that violence is widespread in society than people who are less exposed to them and see the world as a scary and unsafe place where people are mostly selfish and untrustworthy. This feeling or idea may encourage the public to tolerate violence and believe that violence is uncontrollable, and people must use violence in order to survive or solve problems. Second, on television violence is rarely punished, instead it is a heroic performance. Third, violence is an effective way to resolve conflicts and achieve personal goals. Other conflict resolution methods such as compromise, debate and reconciliation are often ineffective. In these situations children are more likely to imitate television violence to resolve conflicts in their lives. They are llusions that we have made after watching violence programs for a long time (Yunxiao & Wei, 1998).

(2) *The influences of media credibility on social perception are examined.* The empirical studies show that media credibility is related to audiences' perception of media.

Taking mass media credibility as a variable that affects audiences' perception, empirical research conducted by some scholars shows that under different media coverage of reality and with different personal experience, the audience will make different evaluations of media credibility, which affects their views on the authenticity of the report. The credibility of the two different media, newspapers and television, is not the same as the perception of the authenticity of the report. In the data analysis of the "Survey of the Social Change in the Taiwan Region: the Mass Communication Team" conducted in July 1998, Wang Xu et al. found that in terms of the influences of credibility on social perception, compared with the fact the news report is worse than the actual situation in the perception of "industrial transfer", the respondents with personal experience have a higher appraisal of the credibility of newspapers and it is more likely to regard the media authenticity as social reality. Here the role of TV credibility is not significant. Compared with the fact that the actual situation is more serious than the news report, respondents make a higher evaluation of the credibility of the newspaper regardless of their personal experiences. It is more likely to consider the media authenticity as social reality. The positive correlation between TV credibility and perception of social reality exist only among respondents without firsthand experience. In terms of perception of "violent crimes", compared with the fact that news reports are more serious than the actual situation, respondents who have no personal experience make higher evaluation of the credibility of newspapers, and it is more likely for them to take the media as social reality, The role of TV credibility is not significant; Compared with the fact that the actual situation is more serious than news reports, no matter whether the respondents have personal experience or not, there is no significant correlation between the credibility of newspapers or television news and perception of social reality (Xu et al., 1999).

In a survey of Macao residents, Yufeng of the University of Macao found that the credibility of the local media affected local social construction. The respondents among Macao residents who are more confident in Macao's local media tend to be more optimistic about Macao's prospects. Those who are more confident in the media of Hong Kong are more pessimistic about Macao's prospects (Yufeng, 2004).

5.5.4 Media Credibility, Culture and Social Capital

This part is to further examine the influence of media credibility on the cultural function from the perspective of social capital. The relationship between three variables (credibility, culture and social capital) is discussed. According to Fukuyama, culture determines trust, and trust constitutes social capital. On the one hand, media credibility, like trust, is influenced by culture. On the other hand, it shapes culture. In this way there are two logics: one is culture—media credibility—social capital. Culture affects media credibility and media credibility is a part of social capital; The second is media credibility—culture—trust—social capital. Media credibility affects the shaping of culture, and culture in turn affects the trust of the entire society, and ultimately affects the entire social capital.

5.5.4.1 Social Capital

The term "social capital" was coined by a French sociologist Pierre Bourdieu in 1980. Later, Robert D. Putnam of Harvard University mentioned the concept of social capital in the book *Making Democracy Work* in 1993 and in a series of widely influential articles in *American Outlook*. Then, this concept receives widespread attention and becomes an important sociological concept, and even later an important concept in political science and a widely used interdisciplinary concept. Its content has undergone major changes. At present, although there is no authoritative definition of this concept, it is widely agreed with Putnam's definition as "networks, norms, and trust—that enable participants to work together more effectively to pursue shared interests". The different views on these three aspects constitutes different interpretations of social capital (Huibin, 2000: 6).

Fukuyama discusses social capital from a cultural perspective. He thinks that there are three kinds of capital, i.e. economic capital, human capital and social capital. Social capital is a capability that arises from the prevalence of trust in a society. One of the most important lessons we can learn from an examination of economic life is that a nation's well-being, as well as its ability to compete, is conditioned by a single, pervasive cultural characteristic, that is the level of trust inherent in the society.

Fukuyama argues that with the convergence of systems in today's world, the main factor determining economic competitiveness is the degree of social trust and cooperation constructed by culture, that is, cultural factors are major determinants of social and economic differences (Fukuyama, 2000: 8). In this way the relationship between media credibility and social capital is divided into two aspects: First, media credibility itself is a component of social capital; Second, credibility affects social capital through the inheritance and shaping of culture.

5.5.4.2 Media Credibility is an Integral Part of Social Capital

According to Fukuyama, social capital is the degree of trust in society itself. Then, media credibility as a category of trust relationship is naturally an important part of social capital. Especially in modern society, media credibility, as a system of trust, is one of the characteristics of trust, and it has more significance for social capital.

The impact of social capital on the national economy can be seen from Fukuyama's discussion Regarding the impact of social capital of trust on the development of information technology, Fukuyama thinks that it is far from clear that the information revolution makes large, hierarchical organizations obsolete or that spontaneous community will emerge once hierarchy has been undermined. Since community depends on trust, and trust in turn is culturally determined, it follows that spontaneous community will emerge in differing degrees in different cultures. The ability of companies to move from large hierarchies to flexible networks of smaller firms will depend, in other words, on the degree of trust and social capital present in the broader society. A high-trust society like Japan created networks well before the information revolution got into high gear; a low-trust society may never be able to take advantage of the conveniences that information technology offers.

Fukuyama attributes the decisive factor of a country's competitiveness to the social capital brought about by differences in the culture of trust. Although it amplifies the role of social capital, a number of his arguments are worthy of further discussion.

Fukuyama provides an important perspective, that is, the impact of social capital brought about by cultural differences on our society is a realistic problem. When thinking about material type of capital, we cannot ignore the existence of social capital, which can have the same effect on the social development (Fukuyama, 2000: 29–30).

5.5.4.3 Media Credibility Affects Social Capital Through Its Role in the Inheritance and Shaping of Culture

According to Fukuyama, cultural differences determine the degree of trust within the society, which constitutes the social capital. Culture is a final determinant of social capital. In addition to being a component of social capital, media credibility has an impact on the culture that determines social capital—influencing the functions of cultural heritage and shaping.

Media credibility affects the cultural heritage of the media. The main contents of the cultural heritage of elites, political and economic media that focus on social responsibility are social norms, moral concepts, literature and art. If these media have low credibility, it may leave space for tabloids and small magazines that do not take social responsibility and rely on vulgar low-level content to attract the audience. Undoubtedly this is not conducive to the inheritance of traditional culture, which in turn affects social capital.

Media credibility influences the shaping function of mass culture. In addition to traditional culture, the social culture should have a mass culture that is developing with mass media. The cultural meaning behind trust should not be just the traditional part, but also the emerging and developing mass culture. Therefore, from a cultural perspective, social capital is not static, but varying across space and time.

As for the shaping of mass culture, like the inheritance of culture, the media that emphasize responsibility need to have higher credibility. The mass culture shaped by them is healthy and conducive to social progress. If credibility of the media that focus on responsibility is low, some media with a poor sense of social responsibility may strengthen the shaping of mass culture, which will inevitably affect mass culture, and then influence the social capital.

Therefore, media credibility is an important variable in the composition of modern society, which influences social capital.

Chapter 6
Conclusion

Research on mass media credibility has been done in the United States for more than 80 years. It has become an important research area in communication. The research of American communication scholars focuses on four aspects, i.e. the measurement of media credibility, the analysis of dimensions, the analysis of influencing factors and the impact on the communication effect. However, the research conducted by American scholars is mainly based on empirical study at the micro level and lacks a systematic analysis at the theoretical level. In China the issue of media credibility is a new proposition. There is almost a blank in definition of key concepts, establishment of research paradigms and formulation of theoretical hypotheses.

This book explores the issue of media credibility and defines basic concepts of media credibility in Chinese and English, as well as the relationship between related concepts. It is the first time to discuss the judgment dimensions and factors influencing media credibility from a horizontal perspective of cross-regional differences, and a vertical perspective of social changes. It proposes some new ideas about the analytic theories based on Western society. Originally it put forward to the theoretical hypotheses and generative model of media credibility, comprehensively analyzing impacts of credibility on media functions.

This book discusses five aspects of media credibility theory, i.e. the definition of key concepts, the compositional dimensions, the analysis of influencing factors, the generative mechanism and the impact on effect of communication.

It is found through horizontal comparison with the United States and other places that media credibility in China is high. Further empirical research shows that the meanings behind media credibility in China and the United States are different, i.e. the dimensions of the public judgment whether the media are credible are different. The dimensions of media credibility judgment in China are authoritative and available, while the dimensions of judgment in the United States are factual, fair and accurate. The comparative analysis of differences in the dimensions of judgment finds that they are more related to the political factors of social systems. Simply put, the media in China is integrated with the Party and the government, while the media and the government in the United States are separated on the surface, and there is a difference

G. Yu, *Research on the Communication Effects and Mass Media Credibility in China*,
https://doi.org/10.1007/978-981-19-6242-4_6

Part II
Study on Measurement Approaches of Mass Media Credibility

Chapter 7
Literature Review of Measurement Approaches

Research on media credibility has been done in the United States for decades, and it has accumulated a wealth of results. As a systematic and groundbreaking study on media credibility in China, it is necessary to review previous research and learn their valuable achievements. Through literature review we can explore areas that need to be improved in the previous research to establish our own research theme and ideas.

The research on media credibility originated in the United States has an outstanding characteristic of pragmatism. The actual needs in the process of social development are the driving force and orientation of media credibility research. Therefore, clarifying the relationship between the development of credibility research and the social and historical process will help us grasp the essence of credibility research and avoid superficial research in China. One of the goals for this research is to lay a foundation for exploring theoretical implications of media credibility. From the literature review we must focus on results of credibility research with cultural specialty instead of blindly citing and referencing.

In addition, the focus of this research is to develop media credibility assessment scale. It is essential to conduct an in-depth and comprehensive literature review on research of media credibility assessment in this section and to obtain valuable research results, ideas, methods, etc.

China's mainland, Hong Kong and Taiwan, have studied media credibility. With the Chinese cultural context, it is necessary to introduce research in these three regions separately.

The literature review is divided into three parts, i.e., social and historical processes of research on credibility, development of the media credibility assessment scale and the status quo of research on China's media credibility.

G. Yu, *Research on the Communication Effects and Mass Media Credibility in China*,
https://doi.org/10.1007/978-981-19-6242-4_7

7.1 Social and Historical Processes of Research on Credibility

Credibility research in the field of communication first attracted attention as a variable of communication effect, and then developed into a relatively independent research topic. Since most of the research is conducted in the United States, its development is closely related to the American society and history. Three major changes in the history of the United States made a considerable impact on communication studies. With the Second World War, the advent and popularity of television since the 1960s, and the crisis of media trust in the 1980s, there have been three climaxes of credibility studies. After these three stages of research, credibility studies begun to move from being fragmented to being systematic and to have a more unified perception of basic issues such as the meaning of credibility and the division of research areas.

Recently, the focus of research has shifted from basic research to deeper and broader studies. On the whole the credibility research is still in its infancy. While some consensus has been reached in the American cultural context, some basic issues are still inconclusive.

The review of credibility research achievements is based on the time span or the research categories (such as "comparison of different media credibility, exploration of credibility orientation and analysis of factors of credibility" (Lo & Chen, 1993), or divided into "source credibility and media credibility" (Kiousis, 2001: 381–403). According to the time-spanned literature it is easy to make ambiguities in concepts and research categories, but following the achievements based on the categories it is not easy to understand the relationship between research development and historical changes and the evolution of concepts.

There are three stages in the basic research of credibility. Each stage is represented by the most influential research projects. The emphases and orientations of these research projects are different and have an important impact on the differentiation of credibility research. The following is to review and discuss the research achievements of credibility from the perspectives of time and category, based on the three-phase representative research.

7.1.1 Research on "Source Credibility" During the World War II

Credibility research started with the effect of communication, and early research on credibility focused on "whether characteristics of the credibility of communicator will affect the effect of communication". Credibility is mainly addressed as a sub-topic of communication effect research or a variable. With the development of research, contents are deepened and enriched, and credibility has become a special topic. From its origin credibility is closely related to the communication effect.

7.1.1.1 Overview of Early Research on Source Credibility

The research on the source credibility can be traced back to the fourth century BC. In *Rhetoric* Aristotle pointed out that there are three modes of persuasion: ethos (ethical appeal or credibility), pathos (emotional appeal), and logos (logical appeal). The most effective means is persuasion made by the speaker's quality and character, i.e. ethos, which has the same meaning as the "credibility" we are discussing today. In Aristotle's view, credibility covers "moral character" and "expertise and experience", and credibility depends on whether the speaker has "good sense", "good moral character" and "good will". Good sense refers to the speaker's judgment on the topic being discussed. Good moral character is the personal moral character of the speaker. Goodwill refers to the speaker's attitude towards the audience. Any speaker who is deemed to have these three qualities will certainly gain the audience's trust (Gong, 1994).

During the World War II, issues such as the relationship between the credibility of message sources and the effect of propaganda attracted attention. The Yale School, led by Carl Hovland, conducted a great deal of research on the credibility and communication effects of the "sender" in the "Research on Communication and Attitude Change Program". The research focused on "communication effect". Its approach mainly uses experiments. In the same content of communication, communicators with high reliability are more persuasive than low-reliability communicators. This is consistent with the focus of ancient rhetoric scholars on credibility (its essence can be said to verify Aristotle's view with quantitative research methods), but messengers, unlike ancient people, mainly send messages through live lectures, and use modern media such as films or leaflets. Hovland's credibility research actually covered the communicator and communication channels such as films and prints, but the latter was not the focus of the research at that time. His research did not distinguish between source credibility and media credibility. The study drew two main conclusions. First, credibility is mainly attributed to characteristics of the sender: expertise and trustworthiness. Second, the communication channel itself has some impact on credibility. The first conclusion is consistent with Aristotle's conclusion over two thousand years ago that credibility contains "moral character" and "expertise and experience".

Weiss, Hovland's colleague, believed that the study of credibility needed to differentiate the "internal source" (such as the producer of the message) and the "external source" (such as the media of communication). Following this idea some researchers later referred to Hovland's studies and other research that focused on sources only as "source credibility" studies. Research on media channels or agencies falls into the category of "medium credibility" (Kiousis, 2001: 381–403).

In addition to exploring the relationship between credibility and the communication effect, the research on content of source credibility explores the composition of credibility dimensions. The general research methodology consists of interviews or open-ended questionnaires. Participants are required to describe what they consider to be trustworthy persons or institution characteristics. The vocabulary of characteristic description is collected to construct a semantic differential scale. Through factor analysis the constituent dimensions of credibility are isolated.

Berlo et al. (1969: 563–576) collected 35 credibility index, and through factor analysis they isolated three dimensions, i.e., safety, qualification and dynamism.

In Singletary's research 90 undergraduate students in Journalism and other majors of Virginia Commonwealth University were asked to imagine news source who was, to the individual, the most credible or believable of any, and to write all words they could imagine which "made" the news person credible. Their responses were alphabetized and duplications were eliminated. The completed list of words relating to credibility numbered 203. The list of words was presented in the form of a series of semantic differential type scales. Participants indicated the number "6" if a particular word was "extremely consistent" with the participant's meaning for the term 'credibility", "5" if "very consistent" and so-on. Then, the alphabetized list of words was presented to another set of 181 students. Respondents were about evenly divided between male–female, mass communication or non-mass communication major and upper-lower division standing. According to factor analysis, the first six factors accounted for 48% of the total variance. and the next 10 factors accounted for only 13.8% of the total variance. Since factor loadings after the sixth factor tended to be very low, analysis was limited to the initial six factors i.e., knowledgeability, attraction, trustworthiness, articulation, hostility, and stability (Singletary, 1976: 316–319).

7.1.1.2 Discussion

In the early discussion of the dimensions of credibility there were defects since the number of samples was small, the statistical analysis methods and tools were not perfect. Singletary's research used factor analysis to explore the dimension of credibility. There were only 118 samples of 203 analysis variables. The reliability of the research results is questionable.

Prior to the 1980s the research on the dimensions of credibility (whether "source credibility" or "media credibility") was limited by the scale of the research and had little influence. Although the name of each research subject is "credibility", there are some differences in their actual research. Some research subjects are the original senders of the message, some are the message senders, and some are communication channels, and so the research conclusion about the dimensions of credibility are naturally different. In the 1980s, with the launch of several large-scale credibility surveys funded by mass media organizations, the discussion of dimensions of credibility began to extend and reach agreement. Their research made tremendous progress.

In addition, the research division between "source credibility" and "medium credibility" is the summary of late scholars. But the distinction is not very clear in the early research. Research on media credibility often uses the word "source" to refer to the media in general. Later Kiousis (2001: 381–403) summarizes earlier research, arguing that the research subject of source credibility is mainly the characteristics of the sender or senders of the message (including individuals, groups and organizations). Media credibility research has focused more on the channel.

From another point of view, mass media in modern society is not simply a channel for transmitting messages. It has a set of information processing mechanisms, and selects and processes the collected original information. Each medium has its own characteristics in terms of focus, collection and processing of information. Compared with the original information, the information produced by different media has unique characteristics. "Media" in media credibility does not simply refer to "channel" but to a complex whole which has a function of both source and channel characteristics. Over time this kind of complex whole has actually become a sender of a message with a humane image in the eye of the audience.

In modern society the media increasingly use anonymous messages to protect the source of news, and the competition in the media market has prompted the media to pay more attention to personalized news processing. In this context, to mass media messages people focus on from the original source of the message to "what media publishing the message." In other words, mass media is itself a source of messages in many cases.

The essential difference between the research on "source credibility" and "media credibility" in the United States is not the difference in the research subjects but in the research focuses. "Source credibility" uses the trust between the communicator and the recipient as a tool to achieve the purpose of communication and change the attitude of the recipient. The research focuses on how to establish credibility so as to achieve better effective persuasion, which lies in the persuasive effects of communication. The research on "media credibility" is based on the trust relationship between the media and the public. The focus of the research is to explore what factors affect this relationship and to analyze the trust status of the media from different perspectives. It is quite different from the research of "source credibility" that achieves the purpose of changing attitudes and creating trust (Dube, 1998).

It is easy to understand the subtle distinction when it comes to the American libertarian press. The liberal concept of journalism believes that mass media plays the role of "neutral gatekeeper", and should maintain a neutral and objective stand on message dissemination, but it must not be a propaganda and inculcator of opinions. No matter how much mass media actually affects people in society, or whether mass media plays the role of persuasive tool, it is against basic values of the media that the public and academia of the United States perceive if mass media is studied as a subjective persuasive tool. However, studies of persuasive communication activities conducted by individuals (such as presidential candidates) or organizations (such as environmental protection organizations) are acceptable. Related research should be attributed to the subject areas such as "persuasion", "rhetoric" or "public relations" instead of "journalism". This context of news values should be the most fundamental reasons for American academics to divide credibility research into two research categories: "source credibility" and "media credibility".

Unlike the United States, China has never denied its role as a persuasive tool for mass media. Therefore, in the study of media credibility in China, credibility and the persuading effects of media are not issues to be avoided but the focus of policy research. In the discussion of the issues related to media in China there are some key questions: What is the persuading effect of media in China? What is the relationship between the persuasive effect and media credibility? Whether mass media as a persuasive tool damages media credibility? What is the theoretical relationship between credibility and media theories (authoritarian theory, classical liberalism, media responsibility theory, Marxist media theory, etc.)? It is not conducive to the research if we follow the American research tradition, avoid the persuading effects of media and divide our research into two areas, such as "source credibility" and "media credibility". Thus, in our research we should understand the tradition of American credibility research and avoid confusion of concepts. We should be flexible and don't have to be formalists.

7.1.2 *A Comparative Study of Media Credibility in the 1960s and 1980s*

Television, as a kind of news media, has become mass communication, which causes a series of reactions. The competition between newspapers, as a representative of traditional media, and television has become the focus of media industry. Correspondingly, in the field of communication there occurs the craze for comparative research on media credibility. The related research has enriched and developed credibility research itself. The comparison of credibility has shifted the research perspective from the early stage of communication effect to "media credibility" itself. The study of credibility is independent of the study of communication effect, bringing major progress to the credibility research. Research boom has produced great achievements and pushed people to explore the issue of media credibility from a broader perspective. Studies in this period have laid the foundation for the expansion of media credibility research.

7.1.2.1 Overview of Comparative Research on Media Credibility

Since 1959 the Roper Organization was commissioned by the Television Information Office to conduct a series of national surveys, which lasted more than 20 years. The purpose was to compare the credibility of several major mass media (newspapers, television, radio, magazines, etc.). Roper research indicated the logical necessity of shifting the research focus from the political field to the economic field after the end of the World War II, and the real cause is that TV as a new medium competes with other mass media, especially newspapers.

The way Roper compares media credibility is to ask respondents similar questions:

> If you got conflicting or different reports of the same news story from radio, television, the magazines, and the newspapers, which of the four versions would you be most inclined to believe—the one on radio or television or magazines or newspapers?

According to Roper Organization's media research, television is the most believed medium. In 1959 the newspaper was selected by 32% and television by 29%; in 1964 newspapers was 24% and television 36%; in 1974 newspapers 22% and televisions 51%; since 1984 the newspaper and television selection ratios have stabilized around 24% and 53% (Edelstein et al., 1989).

The comparison method that Roper Organization designed to select the "most believable" medium among several media was later called the "relative believability" measurement. Its advantages are simple and clear, easy to operate. But it does not measure the actual credibility of the media but the selection ratio of each media being the "most trusted" media.

The Roper Organization research was funded by agencies that represented the interests of the television industry, and the design was considered to be biased toward television. The research design of the Roper Organization has been questioned and revised.

One way to modify is not to change the comparison method of "choose one among several" but to set some "cognitive reference points" for comparison. For example, Abel and Worth (1977: 371–375) think that TV and newspaper are compared in general. It is very likely for respondents to compare between the local newspaper and the national television network. Therefore, it is not good for the newspaper. It is necessary to revise the Roper Organization methods of comparing several media channels in general, so that respondents only compare credibility between the "local television" news and the "local newspaper" news. When the local newspaper and local TV news give conflicting or different reports of the same story, respondents are more likely to believe the newspaper or television. Local TV news is still considered the most credible even if the respondent's cognitive referent point is the local news.

Another way to modify is to abandon the "choose one" model and let respondents evaluate the credibility of each medium separately and then compare them. Carter and Greenberg (1965: 29–34) designed questions for respondents:

> We would like your opinion on the reliability of (e.g., radio) for news. If perfect reliability is 100%, in your opinion, what percentage of the news on newspapers (television, magazine, etc.) do you believe (from 0 to 100%).

Studies proved that Carter and Greenberg's way of questioning reduced the gap of reliability between television and newspapers. The average reliability of news on television was given as 82% with the newspapers as 68%. The comparison of credibility in each medium during different periods is known as the "absolute credibility".

Various revised media comparison methods narrowed the credibility gap between television and newspapers, but most studies showed that television was perceived as the more believable medium.

In addition to the comparison of media types, some studies compared the credibility of media or journalists with other social institutions or practitioners (such as government, courts, religious institutions or lawyers, doctors, police, etc.). Such comparisons placed the media in a wider context of social trust, which provided a macro reference standard for the trust of the media.

Roper Organization research caused a craze for comparative research on television and newspaper credibility. The comparative research on media credibility became an important subject of credibility research. Sometimes these studies are too particular about details, even to the point of "picking out" (Edelstein et al., 1989). For example, Shaw thinks that television is more popular and people have more reliance on it. The Roper Organization question is "the most believable medium" following the first question "the most reliable medium, which made more respondents select television due to habitual preference. This actually reflects that the comparative research on media credibility has no room for development. The expansion of credibility research must seek other research directions.

7.1.2.2 Discussion

(1) Accuracy and precision of media credibility measurement

Reviewing the history of comparative studies on credibility, Carter and Greenberg used rating percentage to compare media credibility, largely in order to maintain research consistency and comparability with the Roper Organization. Although the measurement results obtained by Carter and Greenberg's questions and the Roper question is several percentage values, the actual implication is quite different. As discussed above, the Roper question compares the "most believable selection ratio" of each medium. The results obtained are not purely media credibility. This question is obviously biased towards higher popular media, which is a comprehensive reflection of variables including popularization, perception, and contact. The comparison between Carter and Greenberg's questions is the credibility evaluation of different media.

From a statistical perspective the two questions are used as different measurement tools for "credibility". The difference between them reflects the dialectical relationship between the "accuracy" and "precision" of the measurement. For example, for a ruler with "meter" as the minimum scale and a ruler with "mm" as the minimum scale for length measurement, using the "meter" as the scale for measurement, the possibility of error is small, but the accuracy is insufficient; using a ruler with "millimeter" as a scale, the possibility of errors is greater, and the measurement error increases, but it is more accurate. Making a choice depends on the nature of the measurement object and the requirements of the measurement.

The Roper question is more accurate because there are only two assessment scales: "most believable" and "non-most believable". Respondents are required to select the one that they trust most, so there may be smaller errors, but the accuracy of the measurement is lower. Carter and Greenberg's question has a rich assessment scale (0–100%), which inevitably increases the difficulty and error rate when respondents answer, thereby increasing measurement errors. The Roper question has greatly improved the accuracy of measurement. Each medium has a corresponding score. Different scores between 0–100% indicate different levels of respondents' trust in the media. If it is a medium that a respondent has not contacted, he or she may not give a score. So when calculating the average score of credibility, it is treated as a missing value instead of 0.

The Roper question is equivalent to the most believable medium being rated as 100%, and the rest of the media, whether it is secondary trust, distrust, or untouched media, are all rated 0 points. Therefore, the Roper question is beneficial to high-popularity media. The popularity of television is generally higher than that of newspapers. It is likely that some respondents who choose TV as the "most believable" only watch TV but rarely read newspapers, not because television is more credible than newspapers. Thus the Roper question naturally widens the credibility gap between television and newspapers.

A study carried out by Pan Jiaqing et al. in Taiwan of China in 1990, using the Roper question, found that television led newspapers by 40% with television 71.3% and newspaper 29.1%. Because of the low accuracy of the Roper questions, the measured data is insufficient and further analysis is very limited. Using the method of "rating, and comparing separately" rich data can be obtained, which is conducive to further analysis.

The comparison of media credibility is an issue that many credibility studies discuss, and it is the topic the press like to address in the process of media competition. With the emergence of the Internet as a new type of media in recent years, comparative research on credibility has begun to rise. Although questions used by the Roper Organization has some defects, it is still used due to its simplicity and ease of operation. Yet it requires an objective review of its measurement results.

(2) The levels of media credibility comparison

In addition to objectively treating the Roper question, implications of the credibility comparison studies make us take notice to levels of credibility when conducting credibility comparisons, especially when comparing the credibility of specific medium (such as *CCTV* and *Phoenix Satellite TV*, *People's Daily* and *Beijing Times*) or subdivided media categories (such as legal media, sports media, economic media). When discussing the credibility of online media, Schweiger (2000: 37–59) divided the message sources that affect the credibility into six levels from the top to the base in a pyramid-shaped diagram, which can be used as the reference object for credibility attribution:

The first level is the presenter, such as reporters and anchors.

The second level is the source, such as politicians or scholars interviewed in the news.

The third level is the editorial units. They are complete programs or single new items in television or articles in the print media.

The fourth level refers to the media product, which can be likened to product brand.

The fifth level are subsystems of a media type. For example, there are the subsystems of quality papers vs tabloid press, public vs private newspapers, computer magazines or news magazines.

Finally, the sixth level refers to the general credibility of whole media types like newspapers, television, radio, magazines, web, etc.

It can be supposed that the credibility attributed to one single level strongly interacts with, or based on other levels.

In addition to these six levels, the recipients have an overall evaluation of the media they perceive. At what level media credibility should be compared, we should choose one according to our own research purposes and focus, and avoid confusion about levels.

Strictly speaking, the credibility comparison between the media needs to be consistent with the comparative level. When using a questionnaire, subtle hierarchical division often complicates questions and affects effectiveness of answering and increases the measurement error. The overly cumbersome topic design is more accurate, but respondents may not be patient and make a discerning analysis on such a detailed level. Instead, it is likely to affect the accuracy of the measurement. So in research we should make an appropriate balance in terms of measurement accuracy and precision according to our own research requirements.

The comparative study of media credibility is not a research direction with a theoretical depth. It is mainly used for the description of phenomena and references of measurement methods and techniques. In the process of comparative research on credibility, some scholars jump out of the credibility research model of the Roper Organization, realizing that media credibility is not a concept of a single dimension and exploring credibility from a multi-dimensional perspective. The research direction provides a larger space for the credibility research. A series of surveys conducted by American Society of Newspaper Editors (ASNE) in the mid-1980s inherited and developed this research orientation and made it the mainstream of media credibility research and the basis for further research.

7.1.3 Discussion on Credibility from the Multi-Dimensional Perspective in the 1980s

Since the 1980s there have been media fake news incidents in the United States. *The Washington Post* reporter Janet Cooke has won the Pulitzer Prize for fabricated news. Many investigations have pointed out that media credibility is in a crisis stage. The number of subscribers of newspapers have fallen sharply since the 1970s, and the credibility of the news media continues to cause concern. The discussion of media trust during this period is no longer just a supplementary issue of other research topics (such as communication effects, media competitiveness, etc.) but has become the focus of research. Correspondingly, the research on media credibility is more in-depth and meticulous. From a multi-dimensional perspective the credibility research has become a mainstream trend, and a certain consensus has been reached on basic issues such as credibility connotation and credibility measurement.

7.1.3.1 Overview of Research on Multi-Dimensional Media Credibility

In the context of the "trust crisis" of the American media, four large-scale national surveys of the public on credibility funded by different media agencies in the mid-1980s were launched almost simultaneously. Each of surveys chose a somewhat different approaches to study and used a different definition of credibility. A national survey by *Los Angeles Times* poll concerned perceptions of political bias in newspaper reporting and evaluations of media in performance, accuracy, fairness, responsibility, and role in criticism of government. The Gannett Center conducted studies in two communities, focusing on the role of public knowledge, media use, and personal experience as well as media credibility. *The Times Mirror* study used both cross-sectional and panel national surveys, scrutinizing public knowledge of media, ratings of accuracy and believability, press criticism with regard to press behavior and practice, political bias, independence, freedoms and government criticism (Gaziano & McGrath, 1986: 451–462). Sponsored by ASNE the national survey of the public on newspaper credibility conducted by Gaziano and McGrath made the most extensive impact. The survey examined credibility from a multi-dimensional perspective, and they proposed a set of scales for assessment of media credibility that received widespread attention, which is adopted and verified by many scholars.

The survey summarized some descriptive items about media credibility in the previous credibility measurement research. Based on the results of focus group interviews, they constructed an overall credibility score with 16 items measuring people's attitudes toward a five-point semantic differential scale. Then the researchers performed a factor analysis on the data to find two factors: "credibility" with 12 items and "social concern" with three items. The ASNE study was of particular importance because its results were disseminated not only through the scholarly press, but through professional channels to newspaper editors and other publishing tradespeople (West, 1994: 159–168). The media credibility assessment scales they designed are widely used and verified.

In 1988, based on the national survey data of ASEN and other survey data the Meyer validation study of the Gaziano–McGrath scales proposed a different structure of factors. After deleting several items two factors, i.e. believability and Community affiliation, were isolated. Meyer made further data analysis of the subset of the credibility scales and simplified it into the five- item credibility index, shown in Table 7.1.

Meyer's colleague West (1994) used Structural Equation Modeling (SEM) to conduct across-validation test of Meyer's subscales and the Gaziano–McGrath scales. It was found that the Meyer's credibility subscale had acceptable goodness-of-fit,

Table 7.1 The Meyer subscale of credibility

Fair—unfair
Biased—unbiased
Tell—doesn't tell the whole story
Accurate—inaccurate
Can be trusted—can't be trusted

indicating that the measurement model is acceptable. It has been widely used as a measurement of media credibility in many studies.

ASNE and some scholars including Gaziano, Meyer and West continued their research. Now basic issues of media credibility and the credibility evaluation scale have more consistent results. Based on this, it is likely for American credibility research to go further.

7.1.3.2 Discussion

(1) *Cultural uniqueness of media credibility research*

From the previous literature review we can clearly see that each stage of the American credibility research is closely related to American social development, and it has unique American social and cultural characteristics. Moreover, some scholars pointed out that the criteria for judging media credibility was greatly affected by the cultural and environmental background of the media in the application process, and the meaning of credibility was changing due to cultural differences. This has been proved in a cross-cultural study. Surveys of college students in Germany showed that people judge mass media as "the capacity of the media to solve social problems and build social justice". In Japan people judge and perceive the meaning of media credibility by "the intimacy in the relationship between the media and the public" (Edelstein et al., 1989).

There are cultural differences between Japan, Germany and the United States, but there is no essential difference in social and media systems. China is unique in culture and has huge differences in its social and media systems compared to the West. In addition, the whole Chinese society is in the process of reforms and changes. It can be inferred that the media credibility evaluation in China should not directly use the American credibility scales.

(2) *Media credibility scales to be improved*

Although the credibility subscale developed by Meyer is widely used, from the test data of Meyer's subscale, it is not perfect and fails to reach the ideal standards in the test indexes. The "Social Concern" subscale did not pass the test, which meant that the factors of social relations between the media and the public in media credibility was not reflected on the credibility evaluation scale. Therefore, the Meyer's subscale as a media credibility scale needs to be further improved.

7.1.4 *Research on Factors of Media Credibility*

Since the 1980s the focus of credibility research has expanded with research conducted by Becker, Whitney, and Collins (1980) on the correlation between perceptions of media use and credibility. Izard (1985: 247–255) discussed on how the American public assessed the performance of the media and media practitioners and found that their trust in the media was related to their evaluation of how well the news media live up to their expectations. However, in the case where the meaning of credibility and the assessment scale were not unified, the research tended to focus on exploring the composition of credibility dimensions (related to the construction of the credibility scale). If there is no agreed-upon credibility scales, research lacks of unified foundation.

The research results of Gaziano, Meyer, West et al. laid a foundation for further research on media credibility. Most researchers directly used relevant indexes or modified them slightly to evaluate media credibility.

The scale is used by scholars in research on different subjects, and media credibility research has developed at a broader and deeper level. If the issue of "what is media credibility" is addressed when discussing the composition of credibility dimensions. Then the research develops in two logical directions.

The first is "Why is credibility so?" It explores the causes and factors of credibility;

The second is "What impact does credibility make?" It explores the impact of credibility on communication and society.

The impact of credibility is not the main topic of this research, so the following makes a brief introduction to the factors influencing credibility.

7.1.4.1 Overview of Factors Influencing Media Credibility

The research on factors influencing media credibility can be divided into different perspectives and levels: internal factor research, external factor research and factor research at the macro level.

(1) *Research on internal factors of media credibility*

Starting from the message sender, research were conducted on the influence of the communicator' s behavior and information characteristics on credibility. This research focus is to explore "what are the characteristics of the media that cause people to trust or distrust", such as "factual reporting", "fairness and unbiasedness", "caring for the public interest", etc. The previous discussion on the composition of factors influencing media credibility belongs to this kind of research. It can be called the study on the "internal factors" of media credibility. The internal factors of the media play an important role in media credibility, and this research orientation contributes to the work of the communicator and the credibility of the media. Thus, it has been the most important topic of credibility research.

(2) *Research on the external factors of media credibility*

Starting from the message receiver, the research was conducted on the influence of factors such as the personality or psychological characteristics of the audience on media credibility. It can be called the study on "external factors of media credibility. This issue has been discussed since the 1960s, when the research focused on the basic demographic characteristics of the audience (such as gender, age, education, etc.) affecting credibility evaluation. For example, women more than men make higher evaluation of median credibility. The women trust television instead of newspapers more often than the men. People with a higher education level make lower evaluation of media credibility (Westley & Severin, 1964; Carter & Greenberg, 1965; Abel & Wirth, 1977). Highly educated people believe in newspapers more than television (Ganahl, 1994; Carter & Greenberg, 1965). Later research expanded to explore the more diverse personal characteristics and media use, such as media user's income, media knowledge, trust in media practitioners, audience's character, and media contact, media reliance, media use preferences, etc. Most studies have found that these factors have certain effect on media credibility.

(3) *Research on factors influencing media credibility at the macro level*

The above internal and external factors of media credibility are all direct factors that affect media credibility. At a macro level these direct factors are caused by macro factors such as media systems, society, economy and culture. Such factors can be called "macro factors" of media credibility. This type of research covers a wider range of contents. A discussion of media credibility from the macro perspective of the social environment requires research in other social disciplines such as sociology, ethics and economics. However, such studies have not been conducted extensively in credibility research.

7.1.4.2 Discussion

The research on the factors affecting media credibility has gone from internal factors to external factors, from demographic characteristics and media use to the psychological level of the audience, from the micro factors to the macro factors. The discussion on factors of media credibility is not only an integral part of credibility research, but its related research affects the overall credibility research and triggers a difference in perceptions of media credibility itself. Research on factors of credibility develop the connotations of credibility. It has brought about the expansion of credibility research areas.

7.2 Development of the Media Credibility Assessment Scale

The above is a review of the literature from the perspective of the socio-cultural history of credibility research. It can be seen that the exploration of the credibility assessment scale is important. There is no systematic research on media credibility in China. Therefore, it is necessary to start with the basic issue of the credibility dimensions of China's mass media and the corresponding measurement index. This issue is the primary concern of this section, so the following is literature review of the development of existing credibility scales.

Since the previous section reviews the literature from the perspective of horizontal development, the following may overlap with the previous discussion. It is mainly based on the credibility assessment scale and makes a more in-depth discussion on relevant literature from the statistical analysis techniques and analysis results, which is the focus in this section.

7.2.1 Overview of Research on Credibility Scales

Earlier research results on the credibility evaluation scale were numerous but fragmented. The main method used was to list items relating to credibility, constructing a semantic differential scale or the Likert scale, and then use factor analysis to integrate many items into a few factors as dimensions and measurement indexes of credibility. Researchers have different research purposes and methods. Some research subjects are aimed at speakers, some are targeted at journalists, and some are news media. Therefore, there are many but not agree-upon definitions of credibility index and dimensions. The process of developing credibility measurement indexes and dimensions is the process of using empirical methods to explore the connotation of credibility. Many researchers think that the credibility measurement dimensions and indexes indicate the meaning of credibility.

7.2.1.1 Early Fragmented Research

In an era when mass media have not yet been produced and popularized, the discussion on the composition of credibility is based on the research on the message sender, that is, the research on the "source credibility". Aristotle believes that "Ethos" ("credibility") includes two main factors, i.e. "moral character" and "expertise and experience". This point of view is proved by Hovland through experimental method, who thinks that credibility is mainly attributed to the characteristics of the sender, including expertise and trustworthiness.

In 1968, Markham (1969: 57–64) asked undergraduate students to view kinescopes of three "off the air" local newscasts and evaluate each of the newscasters on a fifty-five item semantic differential instrument, and identified the three major dimensions of source credibility (see Table 7.2).

In 1969 Berlo et al. (1969: 563–576) sorted out 35 items related to credibility based on Markham's research, and the factor analysis isolated three dimensions of credibility (see Table 7.3).

In 1969 Jacobson (1969: 20–22) used factor analysis to divide credibility into four dimensions (see Table 7.4).

In 1973 Shaw (1973: 306–311) developed a set of seven-level semantic differential scales, using five groups of adjectives as the dimensions of credibility (see Table 7.5).

In 1975 McCroskey and Jenson (1975: 169–180) conducted a survey with reference to Markham's scale and performed a factor analysis of 46 semantic differential items to identify different credibility dimensions (see Table 7.6).

In 1976 Singletary's research on the credibility of news sources believed that the reason why the previous research reached different conclusions through factor analysis was that the original items used were different. Therefore, the items related to credibility should be identified first. 90 students in journalism and other majors were asked to describe a news source who was, to the individual, the most credible or believable of any. The new source could be any real person on television, radio, newspaper, magazines, wire service or book. A total of 203 words related to credibility were collected.

Through factor analysis, the first six factors were isolated as six dimensions of the source credibility (Singletary, 1976: 316–319) (see Table 7.7).

In 1977 Abel and Wirth (1977: 371–375) examined the credibility of the media with local news as the research subject, and classified credibility into three dimensions (see Table 7.8).

In 1978 Lee (1978: 282–287) found that the evaluation of media credibility was affected by different media and different types of news. In the research respondents were asked to measure the credibility of newspaper national/international news. The credibility presented four dimensions (see Table 7.9).

When measuring the credibility of TV national/ international news, credibility presented three dimensions (see Table 7.10).

When evaluating local and state news in newspapers, credibility presented the following four dimensions (see Table 7.11).

Table 7.2 The Markham credibility scale of the newscasters

Reliable—logical
Showmanship
Trustworthiness

Table 7.3 Berlo, Lemert, and Mertz credibility dimensions of the sources

Safety
Qualification
Dynamism

Table 7.4 Jacobson credibility dimensions

Authenticity
Objectivity
Dynamism
Respite

Table 7.5 Shaw credibility dimensions

Unreliable—trustworthy
Selfish—public spirited
Fragmented—complete
Biased—impartial
Reckless—prudent

Table 7.6 McCroskey and Jenson credibility dimensions

Competence
Character
Sociability
Composure
Extroversion

Table 7.7 Singletary credibility dimensions of the source

Knowledgeability
Attraction
Trustworthiness
Articulation
Hostility
Stability

Table 7.8 Abel and Wirth credibility dimensions of the source

Believability
Trustfulness
Importance

Table 7.9 Lee's credibility dimensions of newspaper national/international news

Trustworthiness
Intimacy
Expertise
Availability

Table 7.10 Lee's credibility dimensions of TV national/international news

Bias
Intimacy
Dynamism

Table 7.11 Lee' s credibility dimensions of newspaper local/state news

Trustworthiness—authenticity—agreeableness
Dynamism
Intimacy
Bias

Table 7.12 Lee's credibility dimensions of TV local/state news

Trustworthiness—authenticity
Immediacy—intimacy
Dynamism—expertise

Table 7.13 *Times Mirror* credibility dimensions of the press

The special interests and the press
The press and its performance
The character of the press

In the credibility of TV local/ state news, it presented three dimensions (see Table 7.12).

In 1986 the research of *Times Mirror* (1986) paid particular attention to the credibility of the press. After factor analysis, the credibility of the press presented four dimensions (see Table 7.13).

7.2.1.2 The Development of a Systematic Media Credibility Scale

As the research progresses, researchers in different periods made an attempt to modify and simplify the previous credibility index. The research has gradually become systematic.

ASNE commissioned Gaziano and McGrath to conduct a national survey of media credibility through focus group interviews and telephone interviews in 1984, with a poll of 1, 200 adults over 18 years of age. The "Five-level Semantic Differential Scale" was constructed with a set of credible metrics consisting of 16 items. This scale is comprehensive and reasonable, which is adopted and modified by researchers later. The credibility scale developed into a systematic one.

The research questions designed by ASNE are as follows:

> Now, we'd like you to think about the daily newspaper you are the most familiar with. Please circle the number in between each pair (of words and phrases with opposite meanings) that best represents how you feel about the daily newspaper (television news) you have in mind.

The newspaper and TV news were measured separately. The principal components factor analysis (using varimax rotation) yielded a "credibility" dimension with 12 indexes and "social concern" dimension with three indexes. The factor analysis results are shown in Table 7.14 (Meyer, 1988: 567–574).

The resulting factor analysis shows that Gaziano–McGrath obtained three factors. The "patriotic" item in the second factor was deleted because differences in factor analysis of television and newspaper, while the "concerned about community's well-being" item was attributed to the first factor due to the unambiguously loading items on the first factor. It was concluded that two dimensions were "credibility" and "social concern".

In 1988 Meyer conducted further exploration and modification on the indexes of the Gaziano–McGrath credibility scale. He pointed out that the Gaziano–McGrath

Table 7.14 Factor analysis results of the Gaziano–McGrath credibility scale

Item	Factor 1	Factor 2	Factor 3
Credibility			
Can be trusted	0.723	0.256	−0.086
Separates facts from opinion	0.710	0.096	0.017
Factual	0.709	0.346	−0.100
Tells the whole story	0.705	0.168	0.033
Accurate	0.694	0.162	−0.021
Unbiased	0.663	0.127	−0.043
Fair	0.660	0.259	−0.034
Respects people's privacy	0.658	−0.214	−0.020
Concerned mainly about the public interest	0.568	0.432	−0.030
Reporters are well-trained	0.566	0.463	−0.062
Watches out after your interests	0.548	0.318	0.044
Patriotic (deleted)	0.184	0.694	−0.048
Concerned about the community's well-being	0.467	0.594	-0.060
Social concern			
Immoral	−0.090	−0.107	0.767
Does not care what the reader thinks	0.101	−0.060	0.712
Sensationalizes	−0.145	0.469	0.553
Variance	37.3%	9.1%	6.5%

scale was lack of face validity, and the three corresponding indexes of the "social concern" dimension lacked theoretical support. These three indexes were the three reversed items in the 16-item scale. Through multiple empirical data, Meyer confirmed that the reverse-polarity items affected the respondent's response and the resulting factor analysis. Therefore, he concluded that the factor analysis result of Gaziano–McGrath was obtained by reverse-polarity items rather than the meaning of the item itself. In the same survey Meyer conducted a group comparison study. The first group received the same items in random polarity as the Gazaino-McGrath scale. The second group received all items in positive polarity. The third group received the items in half positive-polarity and half negative ones. The resulting factor analysis of the first group was a reasonable approximation of the original. The factor solution of the second group was quite different from the results of the first group. Only two factors emerged, and they were not as clearly differentiated. As many as 11 items that load more than 0.3 on both factors. The results of factor analysis are not consistent with common sense, and it is difficult to make a theoretical explanation. The third group yielded an idiosyncratic four-factor solution, which could be completely unexplained.

Table 7.15 Factor analysis results after deleting the reverse-polarity items

Item	Factor 1	Factor 2
Believability		
Unbiased	0.713	0.158
Tells the whole story	0.679	0.279
Can be trusted	0.673	0.421
Separates facts from opinion	0.668	0.267
Accurate	0.663	0.285
Fair	0.632	0.377
Respects people's privacy	0.618	0.068
Reporters are well-trained	0.438	0.587
Factual	0.587	0.531
Community affiliation		
Patriotic	−0.006	0.741
Concerned about the community's well-being	0.293	0.710
Concerned mainly about the public interest	0.361	0,658
Watches out after your interests	0.334	0,581
Variance	45.6%	7.7%

Furthermore, Meyer analyzed the original data of Gaziano–McGrath, deleted three reverse-polarity items, and yielded the two-factor solution (see Table 7.15 for details).

Two of the 13 items, "factual" and "reporters are well-trained" loaded strongly on both factors, and because of this ambiguity were dropped from further analysis. In this way Meyer's media credibility scale had a total of 11 indexes, of which seven were attributed to the same factor "believability" as Meyer named. The other four indexes were attributed to another factor named as "community affiliation". Reliability tests were performed on the two subscales, and the Alpha coefficient for the "believability" subscale was 0.84. The Alpha coefficient for the "community affiliation" subscale was relatively low at 0.72, but it reached an acceptable level.

Meyer (1988: 567–574) further simplified the "believability" sub-scale. After repeated reliability tests, two items of "respect people's privacy" and "separate facts from opinions" were deleted, and the five-item credibility indexes was finally produced, which were the "unbiased" and "tell the whole story", "can be trusted", "accurate" and "fair". The alpha coefficient is 0.83. Dropping any one would degrade alpha to 0.8 or just below. Meyer proposed that the five remaining items constructed a sound "believability" index, as shown in Table 7.16.

The research results of Gazaino and Meyer have a widespread influence. Not only are they widely used in the academic and industrial circles in the United States, but credibility studies in Taiwan of China are also based on the related research. Zhang (1989), Kuai (1989), Lu (1992), Peng (1991), Lo and Chen (1993) conducted their

Table 7.16 Meyer's "believability" index

Item	Item-total correlation	Alpha if item deleted
Fair—unfair	0.678	0.78
Biased—unbiased	0.609	0.80
Tell—doesn't tell the whole story	0.632	0.79
Accurate—inaccurate	0.609	0.80
Can be trusted—can't be trusted	0.612	0.80

research on credibility measurement and connotation on basis of Gazaino-McGrath and Meyer's credibility index.

In 1993 "Research on the Credibility of the News Media", a research project in Taiwan of China, chaired by Lo Ven-hwei and Chen Shimin, was a large-scale research on media credibility in the Chinese academic community. Following the research results of Gazinao, the study designed a Taiwan region media credibility evaluation scale consisting of 16 items. The questions were designed as follows:

> We would like you to rate the credibility of newspapers and TV news. According to your personal perceptions, please rate the newspapers and TV news you usually read based on the performance of newspapers and TV news in the following items. You can score from 0 to 100, 0 refers to the lowest, 100 points refers to the highest, and 60 points refers to pass. Even if you are not sure how well the newspaper and TV news performed, please give a score based on your perceptions.

(1) *In terms of the credibility of the news*
   a. *How many points will you give to the newspaper news?*
   b. *How many points will you give to the TV news?*

(2) *In terms of social welfare?*
   a. *How many points will you give to the newspaper news?*
   b. *How many points will you give to the TV news?*

(3) *To what extent is the news not exaggerated?*
   a. *How many points will you give to the newspaper news?*
   b. *How many points will you give to the TV news?*

(4) *In defending the public's right to know?*
   a. *How many points will you give to the newspaper news?*
   b. *How many points will you give to the TV news?*

(5) *In terms of unbiasedness of news?*
   a. *How many points will you give to the newspaper news?*
   b. *How many points will you give to the TV news?*

(6) *In promoting social reform?*
   a. *How many points will you give to the newspaper news?*
   b. *How many points will you give to the TV news?*

(7) *In terms of detail and completeness of the news?*
   a. *How many points will you give to the newspaper news?*

b. *How many points will you give to the TV news?*

(8) *In regard to the public opinions?*

a. *How many points will you give to the newspaper news?*

b. *How many points will you give to the TV news?*

(9) *In terms of the factualness of the news?*

a. *How many points will you give to the newspaper news?*

b. *How many points will you give to the TV news?*

(10) *In terms of being concerned about social welfare?*

a. *How many points will you give to the newspaper news?*

b. *How many points will you give to the TV news?*

(11) *In terms of the news reliability?*

a. *How many points will you give to the newspaper news?*

b. *How many points will you give to the TV news?*

(12) *In terms of being concerned about the public interests?*

a. *How many points will you give to the newspaper news?*

b. *How many points will you give to the TV news?*

(13) *How accurate is the news?*

a. *How many points will you give to the newspaper news?*

b. *How many points will you give to the TV news?*

(14) *In respect of national interests of China?*

a. *How many points will you give to the newspaper news?*

b. *How many points will you give to the TV news?*

(15) *How fair is the news?*

a. *How many points will you give to the newspaper news?*

b. *How many points will you give to the TV news?*

(16) *In respecting the public's privacy?*

a. *How many points will you give to the newspaper news?*

b. *How many points will you give to the TV news?*

At the beginning of the scale design, Lo Ven-hwei et al. treated the dimension of credibility as "credibility" and "social concerns" and designed eight corresponding items for each factor, based on the research results of Gaziano–McGrath. Because the Gazaino scale is mixed with the positive-and-negative polarity items, which makes the factor analysis results obviously inconsistent with common sense. For example, the "concerned mainly about the public interest" and "concerned about the community's well-being" group with the "credibility" factor, while the "sensationalizes" groups with the "social concern" factor. Lo Ven-hwei's scale is closer to the factor analysis result obtained by Meyer in 1988 after deleting the reverse-polarity items based on the original data of Gazaino-McGrath. The indexes and dimensions of the three scales are shown in Table 7.17.

Lo Ven-hwei's scale and Meyer's factor analysis results are basically consistent. The difference is in terms of "social concern" factors. There are three additive factors in Lo Ven-hwei's scale including "defends the public right to know", "promotes

Table 7.17 Comparison of three scale indexes

Lo Ven-hwei scale	Gajino-McGrath scale	Meyer scale
(Credibility)		
Fair	Fair—unfair	Fair—unfair
Factual	Factual—opinionated	
	Separate facts from opinions—confuse facts with opinions	Separate facts from opinions—confuse facts with opinions
Detailed and complete	Tell the whole story—do not tell the whole story	Tells the whole story—doesn't tell the whole story
Correct	Accurate—inaccurate	Accurate—inaccurate
Impartial	Unbiased—biased	Unbiased—biased
Not exaggerated	Sensationalizes—doesn't sensationalizes	
Trusted	Can be trusted—can't be trusted	Can be trusted—can't be trusted
Reliability		
	Reporters are well-trained—unexperienced	
(Social concern)		
Respects the public's privacy	Respects people's privacy—invades people's privacy	Respects people's privacy—invades people's privacy
Concerned about the public interest	Concerned mainly about the public interest—concerned about making profits	Concerned mainly about the public interest—concerned about making profits
Values national interests of China		Patriotic-not patriotic
Concerned about social welfare	Concerned about the community's well-being—not concerned about the community's well-being	Concerned about the community's well-being—not concerned about the community's well-being
Respects the public opinions	Does not care what the reader thinks — care what the reader thinks	
Defends the public right to know		
Promotes social reform		

(continued)

Table 7.17 (continued)

Lo Ven-hwei scale	Gajino-McGrath scale	Meyer scale
Upholds social justice		
	Watches out after your interests—doesn't watch out after your interests	Watches out after your interests—doesn't watch out after your interests
	Immorality—moral	

social reform" and "upholds social justice", which indicates that Lo Ven-hwei takes account of the social and cultural differences when designing the scale. It may be that in the context of social changes in Taiwan of China, such as lifting the ban on newspapers, political democratization, and economic development, the public have more expectations for mass media.

The second difference is that "respects people's privacy" groups with the "credibility" factor in Meyer's scale, while Lo Ven-hwei's scale groups "respects the public's privacy" with the "social concern" factor.

The third difference is that Lo Ven-hwei groups the two indexes of "separates facts from opinion" and "factual" in the Gazaino scale with the item of "factual".

In addition, there is a translation problem, the "accurate-inaccurate" item in the scales of Gazaino and Meyer, et al., which is generally translated as "correct-incorrect" in studies of Lo Wen-hwei groups. Thus "correct" has become media credibility indicator corresponding to "accurate-inaccurate" used by the Chinese academic community. But accurate is not correct or right, and the translation "accurate-inaccurate" should be more in line with the original meaning of the word. Although "correct", "accurate" and "right" are synonyms, they have different meanings. "Correct" is a subjective judgment to a greater extent. Different people may have completely opposite "right-wrong" judgments of the same medium. This overly subjective criterion is obviously not suitable as a criterion for judging the credibility of the media. "Accuracy" is a judgment on the extent to which the media report reflects the facts. Relatively, it is a more objective judgment criterion and is more suitable as a criterion for judging media credibility.

The Lo Ven-hwei's scale presented above was designed based on experience and existing research results before the survey was implemented, but Lo et al. (1993) conducted factor analysis on the survey data and identified 16-items factor. The two factors, i.e. "credibility" and "social concern" did not appear. But in 1992 Lu Hongyi used a credibility scale similar to Lo Ven-hwei's scale in his master thesis. Taking high school students as research subject, he found that the credibility of newspapers and TV news presents the two dimensions, i.e. "credibility" and "social care" (Wang et al., 1999).

The major statistical technique used in discussion on the credibility scales is the exploratory factor analysis (EFA). In 1994 West pointed out that exploratory factor analysis, while useful at the initial hypothesis-building stage of scale construction, is unable to confirm the validity and reliability of scales. With the advance of statistical innovations, West (1994: 159–168) uses a statistical technique called Structural Equation Modeling (SEM), which have found widespread application in the development of measurement models and the verification of assessment scales. In particular, they are useful for the validation of additive scales, since this modeling can provide direct tests of the suitability of scales for use as additive scales and of the reliability and empirical validity of scales. The three widely-used scales such as the Gazaino-McGrath scale (see Table 7.14), Meyer's simplified "believability" subset (see Table 7.16) and "Society Concern" subset (see Table 7.17) were verified.

After testing Meyer's "believability" subset has acceptable goodness-of-fit in terms of the overall reliability, and the empirical validity and the measurement model, individual item reliability is somewhat lower. The Gaziano–McGrath credibility scale fares less well. Despite high overall reliability measures, the model has insufficient goodness of fit, indicating that there is a problem with the factoral structure of the scale. The Meyer "community affiliation" subscale has acceptable goodness-of-fit, but its overall reliability and empirical validity measures is below acceptable levels. In West's view it does not do so with sufficient reliability or empirical validity to be recommended for use without further development. (West, 1994: 159–168).

Since West conducted a cross-validation test of the credibility scales, the "believability" subscale of five indexes developed by Meyer performed well and was used in more research. For example, in Wang et al. (1999) and Lo et al. (2002) used this scale in their research on media credibility.

In 1999 the scale for measurement of media credibility used by Wang Xu, a scholar from Taiwan of China, contained 5 indexes. each of which there was a sentence about the news credibility. The interviewees were asked to answer whether they agreed with the item, to evaluate them respectively on a scale of 1 to 5 (from "strongly disagree" to "strongly agree"). The items are shown in Table 7.18.

This scale is basically the same as Meyer's "believability" subscale (including five indexes, i.e. "fair", "unbiased, "tells the whole story", "accurate", "can be trusted"), except for the different statements of the items. The principal components factor analysis and varimax rotation are used to isolate a common factor with 50% of the variance explained.

In 2002 Lo Ven-hwei and Lin Wenqi used Meyer's "believability" subscale as a scale for measurement of media credibility in related research on the credibility of election news. There was some difference in the form of questionnaire design. The respondents were asked separately to evaluate the five media:

Table 7.18 Indexes of Wang Xu's media credibility scale

(A certain medium) When reporting controversial issues, it usually conveys the opinions of people from all walks of life
(A certain medium) tells the whole story, and we can get enough information
(A certain medium) is usually neutral in reporting news
(A certain medium) is usually fair in reporting news
(A certain medium) is usually trustworthy in reporting news

Do you think whether the election news on (newspapers, television, radio, magazines, internet)?

a. can be trusted
b. is biased
c. tells the whole story
d. is accurate
e. is fair

The answer was chosen by the respondents from a five-point scale, of which 1 represents very strongly not "can be trusted" (very "biased", not "tells the whole story", etc.), and 5 represents very strongly "can be trusted (very "unbiased", completely "tells the whole story", etc.), those who answered "don't know" were treated as missing values in analysis.

The factor analysis and validation test found that when measuring the credibility of the election news in the five media, one factor was represented, and its Alpha coefficient reached 0.82 or better.

In above two studies the scale design and factor analysis and validation test results are similar to those of Meyer.

7.2.1.3 The Use of Credibility Scales

Since Gazaino et al. made a great achievement in the credibility scale in 1986, research on media credibility has gradually extended to a broader area from the basic issue of specifically exploring the constituent dimensions of credibility. In these studies, the indexes of the scale for media credibility measurement are not specifically discussed, but the scale is used to measure media credibility, and the measurement result is used as a variable in the study. The scales used in these studies are generally based on the results of the research by Gazaino and Meyer et al. and modified accordingly.

Peng (1991), a scholar from Taiwan of China, conducted a survey of newspaper credibility with the Taiwan region's media directors as interviewees in 1988 and 1990 respectively. With reference to the research by Gazaino et al. a 15-item scale was designed (see Table 7.19).

This research conducted a questionnaire survey of media directors instead of ordinary people. It focused on comparing the media measurement differences between directors of newspaper media and ones of television media. Without factor analysis of the scale it was concluded that the person in charge of the media was more tolerant of the media in which he works, and harder on the other media.

In 1994 Vanta and Hu (1994: 90–98) used the credibility scale in the correlation study of media credibility and agenda-setting effects. The scale included two dimensions: one called "believability" and the other called "affiliation." The specific topic was designed to develop a Likert scale. Respondents were asked if they "strongly

Table 7.19 Peng Yun's newspaper credibility scale

1. The number of news reported in the newspaper is just right
2. Factual and unbiased
3. Balances the reports for different opinions
4. Reports things accurately
5. Respects the public's privacy and does not arbitrarily defame personal reputation
6. Separates pure news from comments
7. Not deliberately sensationalizes
8. Selected news topics worth reporting
9. Once a report is wrong, willing to acknowledge and correct it
10. Fluent in writing
11. Values readers' opinions and rights
12. Fair treatment of candidates from different political parties
13. Concerned about the community's well-being
14. Concerned about the public interest
15. Can be trusted as a whole

Table 7.20 Wanta and Hu scale of credibility

Believability
News press such as newspapers and television news try to manipulate public opinions
News press often fails to get all of the facts straight
News media often doesn't deal fairly with political or social issues
News press doesn't do well in separating facts from opinions
Affiliation
News press is concerned with the community's well-being
News press watches out for your interests
News press is concerned mainly about the public welfare

agree, somewhat agree, are undecided, somewhat disagree, or strongly disagree" with the seven statements about the news media (see Table 7.20).

In 1998 Johnson et al. (1998: 523–540) conducted the credibility research comparing the Internet and traditional sources on political messages, including four indexes, i.e. "believable", "fair", "accurate" and" in depth".

In 2000 Ye (2000) used the credibility scale when discussing the online media credibility, including five indexes, i.e. fair, unbiased, correct, can be trusted and in-depth.

7.2.2 Discussion

From the development of the above credibility scale, we can see that the advance of statistical technology has provided technical support for the development of the scale. The constituent dimensions of media credibility and validity and reliability of the scale are not merely judged by subjective experience, but are analyzed and verified through statistical techniques.

7.2.2.1 Consistency of Credibility Scale Research

Generally, the determination of a scale requires several processes such as item analysis, reliability and validity test. Each process uses different statistical analysis techniques to analyze its structure, and judge the fitness of the scale or the items. Through different statistical methods, each constituent dimension and item in the overall scale yield multiple statistical parameters representing their meanings. These parameters are the main reference for analyzing and modifying the scale during the scale construction.

The American academic community follows academic norms, and different researchers continue their research. Therefore, there has been a continuous verification and development of the scales from Gazaino-McGrth to Meyer and West lasting nearly 10 years, which may avoid duplication of research and improve research efficiency. However, the defect is that such a scale is not a result obtained through a complete research process. The original scale was designed by Gazaino et al. used second-hand data to screen and test the scale. Different researchers have different perceptions, and there arise inconsistencies between research design, item selection, and testing. For example, Meyer's deletion of three reverse-polarity items is based on the omission of items at the beginning of the design. After deletion, two factors with significant validity are obtained, but it seems like a coincidence, lack of strong data support. Moreover, screening and testing of the items on the scale were completed by Meyer and West respectively, but in the development of the scale, the screening of items and the test of validity were often combined. The test outcomes provide an important basis for scale modification, but it is difficult to do so in separate studies. From the test results, each of the scales has defects, but West only pointed out some indexes of the Gazaino or Meyer scale were not good, but couldn't modify them based on these results.

7.2.2.2 The Application of Credibility Scales and Statistical Analysis Techniques

From the perspective of cultural differences, the previous section proposed Meyer's "believability" subscale should draw attention to its appropriateness in cross-cultural research. If we look at the results of the data, Meyer's "believability" subscale is not good enough. When the structural equation model is used to verify the data, the goodness-of-fit index (GFI) close to 1 indicates the degree of goodness of fit. It is generally agreed that the model fits the observed data when it is above 0.9 (Guo, 1999: 351; Ke et al., 2003: 376). The GFI of the Meyer "believability subscale is 0.87 after normalization. And the reliability of a single item is not very high. Therefore, although Meyer's "believability" subscale is relatively good, there is still a room for improvement. However, many current studies (especially those in Taiwan of China) tend to directly use this scale as a scale for measurement of media credibility. There is no problem with Meyer's "believability" subscale as the basis for the development of the scale. But it is not serious to give too high reliance on this scale (especially in cross-cultural research).

In addition, the development of statistical technology and the application of social sciences to statistical technology are a process of continuous development. Thus, the statistical analysis techniques used by Gazaino, Meyer and West are continuously improving. For example, the factor analysis techniques were used in Gazaino's exploration of the dimensions of credibility and deletion of scale indexes. Meyer's amendment to the Gazaino-McGrass scale is mainly based on the reference data such as factor loadings of exploratory factor analyses and Alpha coefficient, etc. West's validation of several influential scales is based on richer parameters, including chi-square values, degrees of freedom, GFI, and correlation coefficient between items and factors, reliability of individual items and overall scale. In goodness-of-fit for all models, in addition to the chi-square values, degrees of freedom, and GFI used by West, the RMR, NFI, RMSEA and other indexes are commonly used as important test parameters.

Research on the credibility of the China's mass media began much later. Fortunately, it has received the attention and funding support of the Ministry of Education and is listed as one of the major research projects. Today statistical technology is more advanced, which provides the possibility for a systematic study of the credibility of China's media. In addition, since the study of media credibility in China is almost blank and there are large cultural differences from the West, it is necessary to start with the scale construction.

7.2.2.3 The Construction of Credibility Scale and the Concept of Credibility

The construction of media credibility scale is closely related to the connotation of credibility. The indexes and dimensions of the credibility scale are not only the measure of credibility, but also reveal the connotation of credibility. Since the credibility research is mainly conducted in the United States, its research inevitably has characteristics of American pragmatism. The research aimed to solve practical problems, focusing on problems related to credibility (such as media competition, trust crisis, etc.). There is lack of theoretical discussions on the nature of credibility. Researchers generally explore the connotation of credibility through empirical methods. With no consensus on the essential meaning of the concept, untargeted operational definitions may come out. There are different definitions in different studies. In order to meet the actual needs of the research, different researches have different purposes and focuses, which has once caused a flood of different operational definitions of credibility. Even in the middle of 1980s, the four large-scale surveys on media credibility conducted at the same time had different interpretations of credibility.

It wasn't until the creation of the Gazaino-McGrath scale that people came to gain a more unified understanding of the connotation of credibility, that is, the credibility contains two dimensions, i.e. credibility and social concern. However, there are differences in this regard. For example, in Taiwan of China, Lu Hongyi used factor scales similar to Gazanio's to obtain two dimensions of "credibility" and "social concern" through factor analysis, but Lo Ven-hwei used similar scales and isolated only one dimension. However, the survey conducted by ASNE in 1999 found that the issue the audience was concerned most about journalism were "accuracy", "the closeness of the media to the community", "biasedness" and "sensationalization" (Zhou, 2004a).

The Gazaino-McGrath scale presented technical flaws in mixing the positive-polarity items and reverse-polarity items (the three reverse-polarity items are related to the "social concern"). When reanalyzing the Gazaino's scale, Meyer deleted these reverse-polarity items, and focused on the items related to "credibility", and produced a simple "believability" subscale with good effect of validation test. Thus, currently in the academia some studies have used a credibility scale that consists of two dimensions, while others used the "believability" subscale as a scale of credibility. So does the concept of credibility contain one or two dimensions? What is the essence of credibility?

Meyer and West suggest that the "believability" subscale is only related to a narrow concept of credibility. West (1994: 159–168) argues that "the development of a reliable and valid set of scales measuring community affiliation should be pursued", when the results of the reliability and validity tests of the "community affiliation" subset were not ideal. Research conducted by Richardson et al. (1988) shows that when the public feels that the media has clear community values and promotes community goals, the public is not only more affirmative of the media value, but also trust them even more. The above research reveals that the understanding of the connotations of credibility should be shifted from a narrow measurement of accuracy to a more comprehensive discussion of individual and social values. And what relationship does the dimension of "social concern" have with the concept of credibility? How should narrow and broad concepts of credibility be identified? They are topics worthy of further discussion.

Moreover, although there are no agreed-upon definition of credibility, the meanings of "believe" and "trust" are basically not controversial. From the face validity of the Meyer's "believability" subscale, the operational definition of media credibility contains the index "can be trusted-can't be trusted". Does this repeat a kind of conceptual analysis of the same term?

The origin of these differences is attributed to the lack theoretical exploration of the connotation of the concept of credibility. Furthermore, insufficient research on credibility theory hasn't developed credibility research into the entire academic system of communication studies. Media credibility is still a separate and practical research topic. Thus, the research on China's media credibility must first explore the theoretical definition of the concept, which is necessary for the subject itself and the prospect of credibility research.

7.3 The Status Quo of Research on China's Media Credibility

7.3.1 Research on Media Credibility in Taiwan and Hong Kong

Media credibility research in the Taiwan region began in the 1970s, and was in line with American research. Many media credibility surveys conducted in Hong Kong used a telephone survey. Most measurement instruments did not use multi-dimensional evaluation scales, but directly asked audiences to rate the specific media. Generally, a scale of 0–10 or 1~10 was used for evaluation. Choosing a 10-point scale allows us to compare the research results obtained in different periods and by different researchers.

"Hong Kong News Media Credibility Evaluation Survey" conducted by The Public Opinion Programme (POP) at the University of Hong Kong has been conducted for many years, and it was conducted almost every two months. The target audience were Cantonese-speaking Hong Kong residents aged 18 or above. The phone number sampling approach was to randomly extract the "seeded" number from the residential phone book, and then produced another set of numbers by adding one minus one and adding two minus two. After mixing, the duplicate numbers were filtered and the final samples were arranged in a random arrangement. The question was designed as:

> Please use a scale of 0–10 to rate the credibility of the news media in Hong Kong, with 10 representing absolutely credible, 5 half-half, and 0 being absolutely not credible. What rating would you give the media?

In addition, Chen Taowen et al., working with the Chinese University of Hong Kong, randomly selected and interviewed residents aged 18 or above on the credibility of Hong Kong's major broadcasters, newspapers, and magazines through telephone, using 1–10 points to rate the media. The results are shown in Table 7.21.

In 2001 Su (2001) from the Chinese University of Hong Kong conducted a questionnaire survey of journalists. The survey covered a wide range of issues, including the evaluation of the credibility of various news presses. It also used a scale of 1–10 points to rate media credibility. In this study the survey results of media "credibility" and media "political orientation" are shown in Table 7.22.

Comparing the credibility research between Taiwan and Hong Kong, the Taiwan region follows the research orientation of the United States, and generally considers credibility as a multi-dimensional concept. The credibility research in Hong Kong is not based on subject of credibility. Media credibility is only a variable under other research topics. Therefore, the specific constituent dimensions of media credibility have not been explored further. Instead, the audiences are asked to directly rate media credibility of the media. The focus of the research is to obtain the public's impression of the media and the overall evaluation data for long-term monitoring

Table 7.21 Evaluation of the credibility of Hong Kong news media by Hong Kong residents in 1997 and 1998

Press and media institutes	1997	1998	Difference
	Average score		
(I) Electronic media			
RTHK	7.21	6.76	−0.50*
TVB	7.04	6.62	−0.42*
Commercial Radio	6.77	6.62	−0.15
Cable TV	6.79	6.56	−0.23
ATV	6.68	6.18	−0.50*
Metro Radio	5.83	5.70	−0.13
(II) Newspaper			
South China Morning Post	7.18	6.58	−0.60*
Ming Pao	7.15	6.55	−0.60*
HK Economic Journal	6.60	6.36	−0.24
Economic daily	6.79	6.31	−0.48*
Standard	7.11	6.23	−0.88*
Sing Tao Daily	6.73	6.07	−0.66*
Oriental Daily News	6.54	5.92	−0.62*
Sing Pao Daily News	6.39	5.82	−0.57*
Apple Daily	6.24	5.67	−0.57*
HK Daily News	6.05	5.48	−0.57*
Tiantian Daily News	5.95	5.18	−0.77*
HK Commercial Daily	5.42	4.74	−0.68*
Wen Wei Po	5.04	4.57	−0.47*
Ta Kung Pao	5.24	4.33	−0.91*
electronic media and newspapers overall average	6.44	5.91	−0.53*
(III) Other media			
Next Weekly	5.57	5.12	−0.15
East Weekly	5.10	4.95	−0.15
Information Services Department	7.34	6.12	−1.22*
Xinhua News Agency HK	5.73	4.94	−0.79*

Note *indicates significances that reach $P < 0.5$. The score range is 1–10

of media credibility, or questions related to other issues (such as press conduct, economic interests, etc.) are used to evaluate the media.

Table 7.22 Media ratings of Hong Kong journalists in 2001

Credibility		Political orientation		Audience orientation	
1. HK Economic Journal	7.63	1. Apple daily	6.51	1. Economic daily	7.03
2. Hong Kong Radio	7.55	2. HK iMail	5.70	2. HK Economic Journal	6.96
3. South China Morning Post	7.47	3. RTHK	5.68	3.South China Morning Post	6.61
4.TVB	7.3	4. Cable TV	5.54	4. Ta Kung Pao	6.43
5. Ming Pao	7.27	5. HK Economic Journal	5.52	5. Wen Wei Po	6.25
6. Cable TV	7.23	6. Sing Tao Daily	5.47	6. Ming Pao	6.16
7. Commercial Radio	7.11	7. South China Morning Post	5.45	7. HK Commercial Daily	5.95
8. HK Economic Times	6.95	8. HK Economic Times	5.41	8. HK iMail	5.95
9. ATV	6.85	9. Commercial Radio HK	5.36	9. RTHK	5.85
10. Sing Tao Daily	6.78	10. Ming Pao	5.26	10. Sing Tao Daily	5.84
11. HK iMail	6.44	11. TVB	5.22	11. Metro Radio	5.79
12. Metro Radio	6.42	12. Sing Pao Daily News	5.15	12. Sing Pao Daily News	5.75
13. Sing Pao Daily News	5.89	13. Metro Radio	5.14	13. TVB	5.57
14. Wen Wei Po	5.52	14. Oriental Daily News	5.11	14. ATV	5.48
15. Ta Kung Pao	5.44	15. HK Daily News	5.06	15. Cable TV	5.12
16. HK Commercial Daily	5.42	16. The Sun	5.02	16. Commercial Radio HK	5.09
17. Oriental Daily News	5.14	17. HK Commercial Daily	4.83	17. The Sun	5.05
18. Apple Daily	5.08	18. Wen Wei Po	4.77	18. HK Daily News	5.04
19. HK Daily News	5.06	19. Ta Kung Pao	4.63	19. Apple Daily	4.78
20. The Sun	4.45	20. ATV	4.54	20. Oriental Daily News	4.44

Note Each score ranges from 1 to 10. Credibility

7.3.2 *Research on Media Credibility in China's Mainland*

The issue of media trust in China's mainland has remained almost blank for a long time. When the SARS occurred in 2003, mass media had a crisis of trust. The issue of the media trust drew the attention of relevant departments of the government. Thus, the academia started to conduct a special research on this issue, which even became a hot topic.

Contrary to the empirical research conducted by the American academic community, most of the discussions on the media trust in China's mainland mainly adopted speculative approaches. With personal perceptions and experiences different researchers proposed different perceptions and opinions of the connotations of media credibility. As in the American academic community, there were various definitions of media credibility due to the lack of empirical and scientific research procedures and testing methods.

The academic research traditions of China's mainland are different from those of the West. There were few reviews and discussions on the research results of their predecessors, especially the Western research achievements. There are inconsistencies in the use of key terms. Some use "believability" and some use "credibility". The meanings of different words are even more confusing. Sometimes different studies use two different words., but their meanings are same. Sometimes the same word is used, but its meaning is different. The differences and correlations between the two concepts of "believability" and "credibility" will be specifically discussed. The following review of literature uses the term "credibility" which is used more frequently to avoid confusion.

Research on media credibility, conducted by scholars in China's mainland, focuses on two aspects: The first is to examine the meaning of media credibility. The second is to explore how media performance reduces or increases media credibility in practice. The following gives an introduction to the research on these two aspects.

7.3.2.1 The Concept of Media Credibility

(1) *Overview of research*

Scholars in China's's mainland define the concept of credibility based on their personal perceptions and understandings. They can be divided into the following categories:

① Viewing media credibility as a way of evaluating and reflecting the trustworthiness of the media.

> Media credibility refers to the degree of the public trust applied to the media (Li, 2003)
>
> Credibility is defined as judgments made by a message recipient (i.e., audience) concerning the believability of a communicator (Wang & Zhao, 2004)

It is the extent to which news media's information products are mainly recognized, trusted and praised by the audience. (Zheng & Tang, 2004)

The credibility of the news media refers to the trust that the news media has among the public. (Li, 2004)

② Viewing the nature of credibility as an ability and determinant to gain public trust and social influence.

Some people believe that media credibility refers to the degree to which the media deserves public trust. So the media need to provide the public with truthful information to prevent unrealism in news reports. The author believes that this is not complete. Media credibility means the power of the media to gain public trust. Media credibility should be shown in the process of the media speaking for the public and helping the public to meet their legitimate requirements. (Wang, 2004)

Media credibility refers to the inherent strength of the media that is trusted by the public. (Chen, 2004)

Media credibility refers to the ability of the news media to win the trust of the public. (Zhang, 2004)

The credibility of the news media refers to the capacities of the news media to gain the trust of the audience. (Zheng & Tang, 2004)

The credibility refers to the power of the media to win the trust of the public. It is the most valuable intrinsic quality of the media and the key factor in the influence and authority of the media. (Cao, 2004)

③ Regarding believability (credibility) as the ability of the media to make influence on the audience. Usually media credibility is consistent with other concepts that express media capabilities such as influence and authority.

Media credibility refers to the credibility, authority and influence that the media has established in society and the audience with its long-term development. (She, 2002)

Credibility is a kind of moral evaluation standard for the media. It is the social power, or social influence, and media capacity out of public trust. Although there is no clear definition, it is generally believed that 'media credibility is the reliability, authority and influence among the audience in the development of media.' As far as the marketing newspaper is concerned, it is the credibility, authority and influence of the newspaper in the audience. (Feng & Yu, 2004)

Credibility is the social power, or social influence, and media power, which is generated by the public trust. (Liu, 2003)

Media credibility refers to a wide range of authority and credibility established over the course of long-term development, which has a profound influence on the audience. (Huang, 1999)

The credibility of the media is that the media provide the audience with authentic, credible, authoritative and noble communication products, and establish trust (honesty, fairness, integrity) and influence among the audience. The greater this trust and influence, the higher the credibility of the media; vice versa. (Li, 2003)

> The credibility of mass media means that the communicator takes social responsibility as their responsibility, and provides authentic, objective, comprehensive, timely and authoritative information through mass media channel so as to obtain social identity. (He, 2004)

(2) *Discussion*

The above interpretations of media credibility seem to have a strong similarity, but actually credibility is defined from three different levels and perspectives.

The first definition is to interpret "力" of credibility as strength and degree, that is, the degree of public trust in the media is credibility, which is the credibility of the media from the perspective of the message receiver. It can be called "degree theory" of credibility. However, this definition is too simple, and the definition still contains other concepts that have yet to be clarified. It is suspected that the same term is repeated: If credibility is said to be "trustworthiness", what is the specific meaning of "trustworthiness"? How is the concept of "trust" defined?

The second definition is to examine credibility from the perspective of the media's own quality, and to interpret the "力" of credibility as "capacity", that is, the "capabilities" of the media that can gain public trust are the credibility of the media. More precisely, "capability" here refers to the "quality" of the medium itself, and this definition can be called "quality theory". Also this kind of definition is too simple, and the specific meaning of "capacity" (quality) of the media is not clear.

The third is to define credibility from the perspective of the function of credibility, and to view the "力" of credibility as "power" and "influence". This definition also refers to "capability" in the same way as the second definition (i.e. "quality theory"). After careful analysis, we can see that the meaning of these two "capabilities" is completely different. The capability in the second definition is the "capability (quality) of the medium", and in the third definition "capability" refers to the "influence" in the audience. It is not difficult to see that this definition confuses the "trustworthiness" of the media with the concepts of "influence" and "authority."

Synthesizing the above three major interpretations of the credibility of the media, we can see that researchers in China's mainland have clear traces of their conceptual meanings based on the Chinese meaning of "credibility". Except that the third definition makes confusion about the concept of credibility, the first two interpretations of "credibility" are consistent with Western scholars' definitions of "credibility". The "degree theory" is similar to interpretation of credibility from the perspective of the audience, i.e. the public attitude and perception of the media. The "quality theory" is similar to the Western view of credibility from the perspective of the communicator, that is, media credibility is the quality that the media itself has to gain trust.

7.3.2.2 Discussion on the Internal Factors of Media Credibility

Most direct factors of media credibility are media performances that can be perceived by the audience and are related to whether the public trust the media. Compared to those external factors such as individual characteristics of the audience that affect media credibility, these factors relating to the media performance can be called internal factors of credibility. Following the academic tradition of serving journalistic practice, scholars in China's mainland focus on this issue. They have started with their own experiences and perceptions, and raised some questions about what media performance reduces or enhances media credibility. Due to different positions and perspectives, some aspects are even contradictory. After reviewing the 17 articles about credibility the different factors are analyzed. The factors' frequency of mention is as follows, from which we can roughly see the degree to which different factors are valued in the academic community, as shown in Table 7.23.

It can be seen from Table 7.23 that the elements of media credibility that researchers propose are basically composed of credibility indexes representing the most important professional quality of media such as "factual", "comprehensive", "unbiased" and "fair", which reflect the similarities between Chinese and Western researchers' interpretations of media credibility.

On the basis of consensus, there are some differences in perception. The two factors of credibility that the researchers in China's mainland most value are "factual" and "close to the people, with humane care". It is not surprising that "factual" is mentioned at most, but "close to the people, with humane care" is so valued, and other constituent elements of credibility valued by Western scholars, i.e. "comprehensive", "unbiased", "fair" are ranked behind "close to the people", "noble style",

Table 7.23 Factors' frequency of mention in academic papers

Internal factors of media credibility	Number of mention	Rate of mention (%)
1. Factual—no false reporting	14	82.4
2. Close to the people and with humane care	11	64.7
3. Noble style—not vulgar, no hype	7	41.2
4. Paid News	7	41.2
5. Public opinion supervision	6	35.3
6. Fair—not biased	5	29.4
7.Complete and comprehensive, satisfying people's right to know	5	29.4
8. Good image	4	23.5
9. Timely and prompt	4	23.5
10. Exclusive, original and independent	4	23.5
11. No false advertising	4	23.5
12. Factual and unbiased	3	17.6
13. follows political correctness and public opinions	3	17.6
14. New and vivid	3	17.6
15. Focuses on major events	3	17.6
16. Avoids hollow propaganda	2	11.8
17. Intermediate price wars and other malicious competitions	1	5.9

"paid news", and "public opinion supervision", which may be an issue with Chinese characteristics. This initially reflects different interpretations of credibility between China and the United States.

Authors of the above 17 articles are academic researchers, and some are practitioners in journalism, or directors of the press. All studies were based on the authors' personal experiences and perceptions instead of the empirical research. Thus, these views are somewhat subjective. Is the media credible? What kind of media is credible? The people who have the most say in these issues should be the public. When determining the constituent elements of credibility, the public needs to identify them, taking the ideas of researchers only as a reference. The best way to collect public opinions is to conduct a social survey.

7.4 Summary

The purpose of reviewing previous research achievements is to learn from them first, and to find out what needs to be improved so as to establish new research themes and research ideas.

Based on this principle, this chapter reviews previous research results, especially the literature related to the connotations of credibility and the construction of the scales. The development and design of the main research topic in this section is based on the literature.

For the sake of brevity, we summarize problems found in the literature review and the research implications and topics in the chart, as shown in Fig. 7.1.

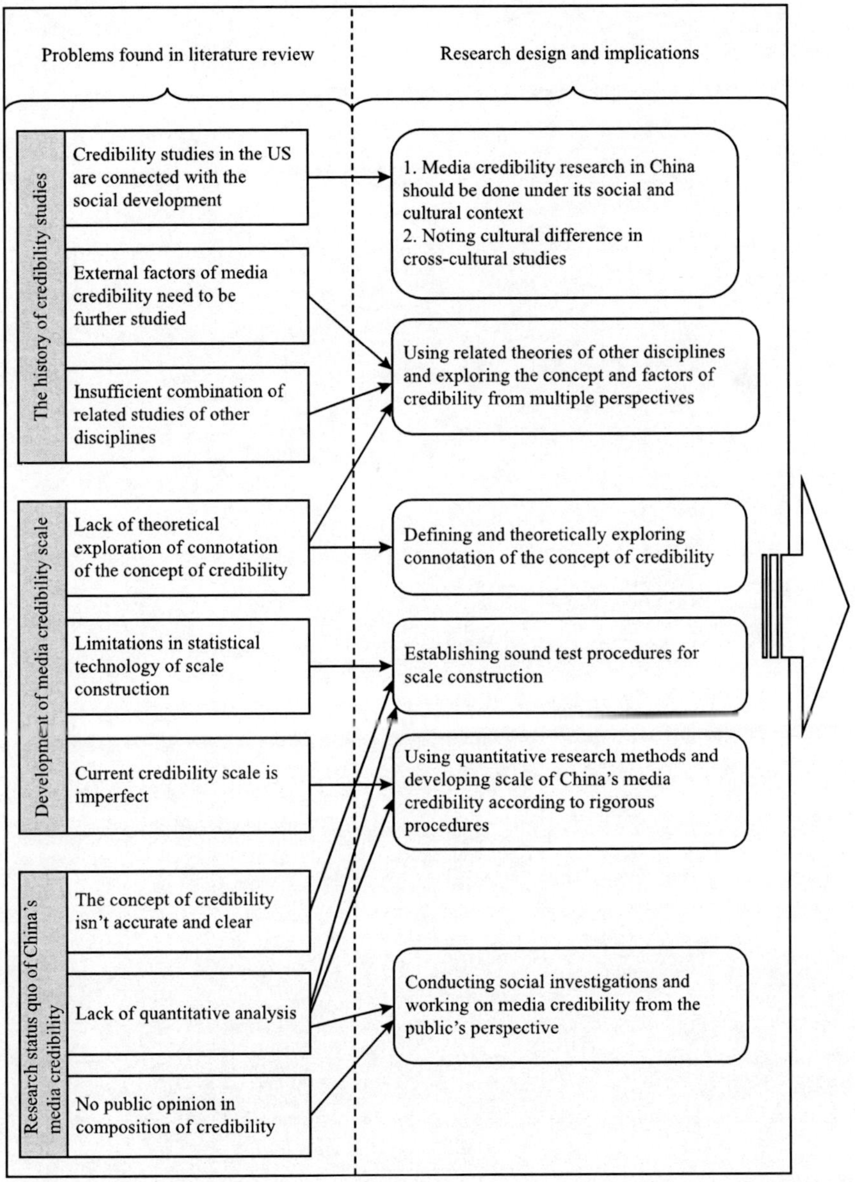

Fig. 7.1 Correlation between literature review and research implications

Chapter 8
Research Propositions and Design of Measurement Approaches

The research propositions have basically two research objectives. The first is to discuss the concept of media credibility. The second is to develop the evaluation indexes of China's mass media credibility, which is the focus of the research. Based on the above research objectives, three research propositions have been established.

8.1 Research Propositions, Methods and Procedures

8.1.1 Research Proposition One

It is to analyze the concept of media credibility (see 1.3.1 of Chap. 1, contractual relationship between the public and mass media). Aiming at the problem that the basic concept of credibility research is not clear, it draws on the research results of "trust" in sociology, economics, and contract theory in institutional economics, etc. to discriminate and analyze the concept of media believability and media credibility from multiple perspectives. Then it proposes the interpretations of the concept of media credibility.

The main research methods used are literature review and speculative method.

The analysis and definitions of the concept are the research starting point and the basis of research ideas and research framework.

8.1.2 Research Proposition Two

It is to investigate the state of Chinese public's expectation for the media, and determining what criteria are used to evaluate media credibility and the importance of different criteria (see Chap. 9).

The main methods used are the focus interview and the questionnaire survey.

G. Yu, *Research on the Communication Effects and Mass Media Credibility in China*,
https://doi.org/10.1007/978-981-19-6242-4_8

The solution to this problem can be divided into the following steps:

Step 1: Through reviewing the previous research achievements and using focus interview or open-ended questionnaires, it collects the feature descriptions of the Chinese people about the media they trust. They are compiled into a preliminary scale consisting of about 30 to 40 items, which is in the form of Likert scale. Each item represents a characteristics of the credible media, called "Chinese Public Media Expectation Scale." Through this scale we will know what "expectation" and "demand" the Chinese people have for the media. The expressions of the "Media Expectation Scale" are obtained by the public on media credibility evaluation criteria with five measuring levels of importance (1 is very unimportant, 2 is not important, 3 is slightly important, 4 is fairly important, 5 is very important). So the scale is also called "Importance Scale of Criteria to Evaluate Credibility" (referred to as "Importance Scale").

Step 2: With the questionnaire survey respondents are asked to rate the importance of each item on the scale.

Step 3: It analyzes the scale data and use statistical techniques such as project analysis, reliability analysis, and factor analysis to describe the areas in which Chinese people judge the credibility of mass media and propose weighted values for different criteria based on the data results.

8.1.3 Research Proposition Three

The public conduct a performance evaluation of China's mass media, according to the results of the importance analysis of the credibility evaluation criteria, and indexes for measuring the credibility of China's mass media are developed (see Chap. 10).

The main research methods used are the focus interview and the questionnaire survey.

The solution to this problem is divided into the following steps:

Step 1: It develops the "Scale for Measuring China's mass Media Credibility". The procedures and contents of the scale are basically the same as the "Importance Scale", except that the "Importance Scale" includes items for evaluating individual media. The "measurement scale" takes the overall Chinese mass media as the object of investigation. The items that are only applicable to the evaluation of individual media are not included. In addition, we use a 0–10 scale for rating each item. The lowest 0 indicates that China's mass media are doing very poorly in this area. The highest 10 points indicate that they are doing very well in this area. A 6-point refers to pass.

Step 2: It uses a questionnaire to ask respondents to rate each item on the scale. (Steps 1 and 2 are performed concurrently with the first two steps of Proposition Two).

Step 3: It uses the exploratory factor analysis and confirmatory factor analysis to analyze the data and follow the analysis results of "criteria importance" to construct a scale for measuring China's mass media credibility.

8.2 Design of Questionnaires, Data Samples and Analytical Tools

8.2.1 *Questionnaire Design*

This research mainly uses questionnaires to collect data and conduct related research and analysis based on the data. The design of the items within the scale is the focus of the questionnaire. In addition to summarizing previous research results, the author conducted interviews on credibility. Some interviews were conducted in the form of face-to-face interviewing, and some interviews were conducted with interviewees through online instant messaging tools (MSN, ICQ, etc.) or through E-mail questioned the author's friends. All the information collected above is a reference for the design of the questionnaire.

Before the formal questionnaire was constructed, the author conducted three small-scale trial surveys on respondents of different ages, identities, and academic knowledge in Renmin University of China, Beijing Library and Beijing Nancheng Newsstand respectively. The questionnaire was improved after each survey. In the trial survey the questionnaire was filled out by the respondents. The author noticed the responses of the respondents in the process of answering and conducted some exchanges with them in order to learn whether the respondents had any objections to the questionnaire design or any problems in understanding questions. Then a brief interview was conducted on media credibility.

After conducting the pilot tests and discussions among members of the research group, the questionnaire design was finally completed. The contents of the questionnaire are not limited to the construction of the measurement scale. Some contents are not covered in this part of research. The design of questionnaire propositions related to the scale construction will be explained later.

8.2.2 *Data Samples and Analytical Tools*

The main purpose of this research is to construct a measurement scale of media credibility, not to evaluate media credibility, so random sampling was not used. Since the scale construction requires to conduct exploratory and confirmatory factor analysis respectively. According to statistical rules, two types of analysis cannot be used for the same set of data, so at least 600 samples are needed.

The survey used a questionnaire. 33 students at Renmin University of China were interviewers after training. A total of 780 questionnaires were distributed, of which 380 questionnaires were conducted in two self-study classrooms on behalf of the liberal arts and science and technology schools at Renmin University of China and Tsinghua University. In addition, 400 questionnaires were conducted in the form of interviews in the areas with a large number of people such as supermarkets and shopping malls in Beijing. Since nearly half of the questionnaires were targeted at college students, the choice of interview sites intentionally avoided the more high-end ones. Small commodity wholesale markets, low-end clothing malls, and popular supermarkets were used as the main visiting locations to obtain samples of people with lower education levels. In order to increase the success rate of the interview, each respondent gave a gift worth about 5 yuan after answering the questionnaire.

Street interviews were conducted on December 11 to 12 of 2004. Surveys at Tsinghua University and Renmin University of China surveys were conducted on December 13 to 14. A total of 732 valid questionnaires were collected. Since this part of research is to construct a credibility measurement scale, 22 questionnaires with a lot of missing values related to the scale construction were deleted in the analysis. Another 18 respondents were younger than 16 years old. Considering the characteristics of media credibility, these 18 questionnaires were deleted. The final number of samples involved in the research analysis was 692, and 361 of them were from the self-study classrooms in Tsinghua University and Renmin University of China, and 331 from street interviews.

The data analysis software mainly employed was SPSS11.0, and for the structural equation model analysis, LISREL8.2 was used. At the same time, Microsoft Excel XP for auxiliary data analysis and chart creation was employed.

8.3 Summary

The logical relationship between the above three propositions constitutes the research contents and context. The three propositions and their correlation are shown in Fig. 8.1.

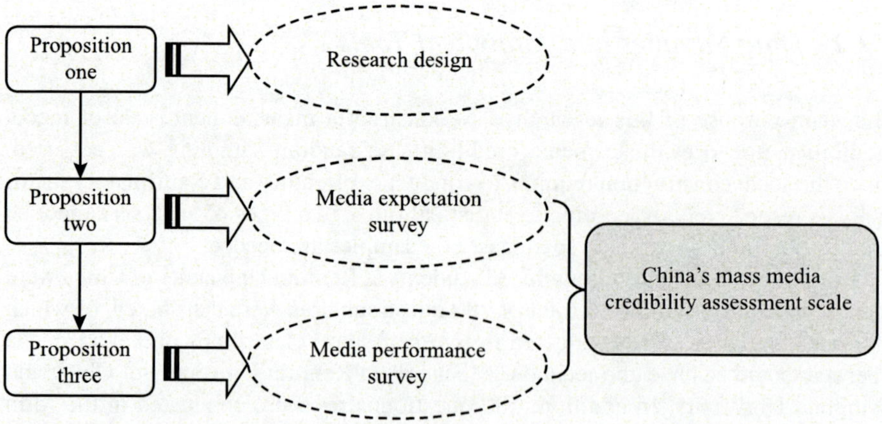

Fig. 8.1 Logic relationship between propositions

Chapter 9
Evaluation Criteria for China's Media Credibility

9.1 Chinese Public's Expectations for the Role of Media

The criteria for judging credibility depend on a role of trustees. For example, the trustworthiness of a lawyer would be considered from his legal knowledge and experience. But people would not care about the legal knowledge when they seek a good doctor. A communicator and an audience are a role partner and they would expect who their partner is (Zheng Xingdong 2005). Therefore, the expectations for the role of media—what role the public think the media should play—have a direct influence on the criteria the public use to determine media credibility.

How do the Chinese people expect the role of media? The answer would affect the fundamental methods to study media credibility. If they have a different understanding of the media's role (i.e. some regard media as the lawyer while the others as the doctor), it will be hard to establish standards that suit for all.

Therefore, before the discussion of the criterion, it is necessary to analyse the public's expectations for the role of media. Based on interviews and research, the initial idea is that most Chinese people have a common criterion in judging the credibility. But some conflicting ideas are noticeable. For instance, some believe "the official newspapers have no credibility at all", while others think "the official newspapers are credible". Whether there is a common expectation for the role of media requires further study and the support of data. In this survey questions are designed as follows:

What role do you mostly expect the media plays? (Multiple choice).

1. A supervisor of the government
2. A mouthpiece of the Party and the government
3. A mouthpiece of the people
4. A platform for free exchange of opinions
5. A tool for entertainment and leisure
6. A tool used by the Party and the government to shape public opinion
7. Others____ (please specify)

G. Yu, *Research on the Communication Effects and Mass Media Credibility in China*,
https://doi.org/10.1007/978-981-19-6242-4_9

Among the seven choices, the most common view is "a supervisor of the government" (67.5%), followed by "a platform for free exchange of opinions" (61.1%), and "a mouthpiece of the people" (54.9%). The three options are over 50%. About a quarter agrees it is "a tool for entertainment and leisure". Less than 15% chose "a tool used by the Party and the government to shape public opinion" and "a mouthpiece of the Party and the government".

In this survey a large part of samples has a bachelor degree or higher, not chosen at random. Therefore, a further analysis based on the education background has been made. We found the significant difference among three groups. For the group lower than a bachelor's degree, less of them chose "a supervisor of the government" and "a platform for exchange of opinions", comparing to the group "with bachelor's degree and higher". For the two groups with higher degree, less of them agree "a tool to shape public opinion" and "a mouthpiece of the government". The smallest difference among three groups is on the option of "a mouthpiece of the people" while the biggest is "a platform for exchange of opinions", nearly 38% between the "undergraduate in college" and "under bachelor's degree". The large difference between the group highly educated and the one with low education level is also found in the "a tool for entertainment and leisure", around 15%, and the former is more affinity to the option. The results are listed in Table 9.1.

In the Western conception of journalism, media is freed from the government control which is the prerequisite for them to supervise the government. Westerners that think the role as "a supervisor of the government" and the role as "a mouthpiece" of government are contradictory. However, in China media is directly managed by the Party and the government, and their supervision becomes "self-oversight". Do Chinese people think the role as "a supervisor" and the role as "a mouthpiece" conflict? Results shows around 13.28% of people who expected the media could be a supervisor chose the role as "a mouthpiece" as well, accounting for 9% of the total sample. 68.9% of people who agreed with the role as "a mouthpiece of the party and the government" also expected the role as "a supervisor". A small number of people

Table 9.1 Perception of the role of the media by the people with different education backgrounds (sample size = 692) Unit: %

Item	A supervisor of the government	A platform for exchange of opinions	A mouthpiece of the people	A tool for entertainment and leisure	A tool to shape public opinion	A mouthpiece of the Party and the government	Others
Overall	67.5	61.1	54.9	25.1	14.0	13.0	3.6
Lower than a bachelor's degree	59.0	36.4	51.6	14.7	16.6	17.1	2.3
Undergraduates	71.9	73.7	57.5	28.9	11.0	10.5	3.9
With a bachelor's degree or higher	70.9	71.3	55.5	30.8	14.6	11.7	4.5

who recognized the role as "a supervisor" chose "a mouthpiece". Among those who agree the role as "a mouthpiece", nearly 70% of them regards it as "a supervisor".

In addition, in Chinese journalism conception, mass media is "a mouthpiece of the Party and the government" as well as "a mouthpiece of the people". Does the public agree with the idea that the two roles are unified?

Results indicates that 17.6% of interviewees who expected the media as "a mouthpiece of the people" chose "a mouthpiece of the government" as well, accounting 9.7% of the sample. Another 8.8% of people chose "a platform for exchange of opinions" and "a tool to shape public opinion". The results are shown in Table 9.2.

Generally speaking, the mostly accepted roles of the media are "a tool to shape public opinion", "a mouthpiece of the people" and "a platform for exchange of opinions". 84.2% chose at least one between "a supervisor of the government" and "a platform for exchange of opinions". 93.9% chose one among "a supervisor of government", "a platform for exchange of opinions" and "a mouthpiece of the people". 21.2% chose one between "a mouthpiece of the Party and the government" and "a tool to shape public opinion", and most of them (85.6%) agreed on another role as "a supervisor" or "a platform for exchange of opinions" or "a mouthpiece of the people". It can be concluded that there is a common view toward the role of the media in the public judgment of media credibility in China although there exist some conflicting opinions that reflect the differing conceptions of journalism in the transformation of the society.

9.2 Design Ideas of the Importance Scale for Evaluation Criteria of China's Media Credibility

9.2.1 Objective I: Developing a Preliminary Set of Indexes for Evaluating Media Credibility

The performances that affect media credibility are summarized based on previous interviews, open questionnaires and literature review. Each would be presented as a form of statement and listed as an item to form "The Chinese Public Media Expectation Scale", also called "The Evaluation Criteria of China's media Credibility Importance Scale" (referred to as the "Importance Scale"). All the items are the basic elements of the "Importance Scale".

The indexes collected are proposed by people with different identities and education backgrounds from different perspectives. They can roughly represent the opinions from following groups: the less-educated people, the intellectuals who are concerned about reality, the researchers in journalism or communications, government administrators and journalists. They have many common views in evaluating media credibility, but they also have different perspectives due to their different situation. What factors are most important to the Chinese public when evaluating media credibility? This is the first objective of compiling the "Importance Scale", which

Table 9.2 Multi-selection of perception of the roles (sample size = 692)

Item		A supervisor of the government	A platform for exchange of opinions	a mouthpiece of the people	A tool for entertainment and leisure	A tool to shape public opinion	A mouthpiece of the Party and the government
Overall	Sample size	467	423	380	174	97	90
	Percentage (%)	67.5	61.1	54.9	25.1	14.0	13.0
A supervisor of the government	Sample size	467	307	283	133	74	62
	Percentage (%)	67.5	44.4	40.9	19.2	10.7	9.0
	Percentage in "a supervisor of the government" (%)	100	65.74	60.60	28.48	15.85	13.28
A platform for exchange of opinions	Sample size	307	423	238	134	61	55
	Percentage (%)	44.4		34.4	19.4	8.8	7.9
	Percentage in "a platform for exchange opinions" (%)	72.6	100	56.3	31.7	14.4	13.0
A mouthpiece of the people	Sample size	283	238	380	109	71	67
	Percentage (%)	40.9	34.4	54.9	15.8	10.3	9.7
	Percentage in "a mouthpiece of the people" (%)	74.5	62.6	100	28.7	18.7	17.6
A tool for entertainment and leisure	Sample size	133	134	109	174	34	35
	Percentage (%)	19.2	19.4	15.8	25.1	4.9	5.1
	Percentage in "a tool for entertainment and leisure" (%)	76.4	77.0	62.6	100	19.5	20.1
A tool to shape public opinion	Sample size	74	61	71	34	97	40
	Percentage (%)	10.7	8.8	10.3	4.9	14.0	5.8
	Percentage in "a tool to shape public opinion" (%)	76.3	62.9	73.2	35.1	100	41.2
A mouthpiece of the Party and the government	Sample size	62	55	67	35	40	90
	Percentage (%)	9.0	7.9	9.7	5.1	5.8	13.0
	Percentage in "a mouthpiece of the Party and the government" (%)	68.9	61.1	74.4	38.9	44.4	100

is to make people establish the evaluation criterion for media credibility by social survey.

The basic idea of compiling the scale is that each item describes a factor that people may consider in evaluating media credibility. All the items are compiled into a scale for social survey. Every item scores from 1 to 5, indicating the importance for evaluating media credibility respectively. 1 "Very Unimportant", 2 "Unimportant", 3 "Slightly Important", 4 "Fairly Important", and 5 "Very Important". By examining the public's views on the importance of each item, the most important ones can be found, which are significant reference for developing indexes or preliminary ones for evaluating media credibility in China.

9.2.2 Objective II: Establishing Reference Values for Media Credibility Evaluation Index

The importance of each index varies among people. The media performance on indexes has different "contribution" to media credibility: if a medium performs good or bad on the factors that are mostly valued by the public, it would have great impact on the credibility. However, the performance on the factors that public perceive as less important will not have a big influence. Here we need to consider how to assign the weight of each index in assessing credibility.

The weight has a direct impact on the measurement results and the correlation analysis between credibility and other variables. There are three main methods for establishing weights. The first one is based on the subjective experience and opinions of the experts, but the opinions of the experts cannot fully and accurately represent the ideas of the people. The second method is more accurate and scientific, which is to establish a causal relationship model including media credibility variables, and then use structural equation model statistical techniques to weight the credibility. However, this measurement needs to establish a model first which requires the support of a large number of theoretical and empirical data. It cannot be achieved now because it needs a long-term research. The third method is the sample survey and the weight of each index could be determined by the Chinese people. From the perspective of actual operability and accuracy, this one is the best choice. Therefore, the second objective of compiling the "Importance Scale" is to provide reference values for the weight of each index.

9.2.3 Choices of the Form of the Scale

There are two major forms of media credibility assessment scales. One is to design two pairs of adjectives or simple sentences with opposite meanings to establish semantic difference scales; the other is to list statements and develop a Likert scale.

The meaning of the items is more clear in sentences. For the first one, proper pairs of antonyms are not easy to find and interviewees may not grasp the exact meaning. Taking into account the education level of Chinese people the items on the scale are presented in the form of statements which could make the meaning clear and easy to understand.

Meanwhile, the "concept + appropriate explanation" is adopted for the concepts that may not be familiar or ambiguous to the general public. For example, the item "Full coverage of news events, not avoiding any important facts" is added a short explanation for the "full", "not avoiding any important facts". The simple explanation will narrow down the meaning of the concept to some extent. Considering that Chinese people are not familiar with the Western journalistic concepts such as "objectivity" and "balance", a clear statement should be more important than the inclusiveness of concepts. Therefore, it is necessary to add a clear explanation to some concepts.

9.3 Design of the Importance Scale

The "Importance Scale" contains 32 items, and the questions in the questionnaire are designed as following,

> When evaluating whether a medium is trustworthy, which aspect would you focus on? Here we list some characteristics of the media. How important are they in your evaluation? Please choose the degree of importance according to your perceptions, using √ on the corresponding □. (1 "Very Unimportant", 2 "Unimportant", 3 "Slightly Important", 4 "Fairly Important", and 5 "Very Important")

The 32 items are roughly divided into 6 sections according to their meanings, which are journalistic professionalism (the inherent requirements of journalistic professionalism), social concern (focusing on the public interest), media ethics (the media is not driven by benefit to harm the public interest), skills in reporting (skills in the presenting events), relationship with the government and external image. The description of the 32 items and the symbols used in analysis are shown in Table 9.3 (the variable symbol is the abbreviation of the variable and the variable label used in the data analysis later).

9.4 Selection of Media Credibility Evaluation Criteria Based on the Importance Scale

The first question proposed by the scale is that which factor that Chinese people value in evaluating media credibility. It is hereinafter referred to as the selection of indexes of importance. It can be answered by the analysis of results.

Table 9.3 Items used in the importance scale and their symbols

Categories	Statement in questionnaire	Symbols in data analysis
Journalistic professionalism	Factual reports, news reports do not contain false, speculative and fictitious elements	EXP01 Factual report
	Full coverage of news events, not avoiding any important facts	EXP02 Full coverage
	Report all news events that the public want to know	EXP03 Complete report
	The news report is particularly accurate	EXP04 Accurate report
	Objectively present the original appearance of the event without adding the bias of the reporter	EXP05 Objective without bias
	For controversial news, balancing the coverage of each party without bias	EXP06 Balancing without bias
Skills in reporting	News reports are prompt and timely	EXP07 Prompt and timely
	Able to provide in-depth analysis and interpretation of complex news events	EXP08 In-depth
	Able to have exclusive news often	EXP09 Exclusive
	The news layout or program is novel, vivid and unique	EXP10 Novel and vivid
	Apologise and correct wrong reports once they have been found	EXP11 Correcting errors
Social concern	The news covers the most realistic and urgent issues in social development	EXP12 Focus on reality
	Good at finding problems and having a sharp perspective	EXP13 Sharp perspective
	Respect for people's privacy	EXP18 Respect for privacy
	With the courage to expose wrongdoing and high proportion of critical reports	EXP19 Exposing wrongdoing
	Concerned about the public interest in their view	EXP20 Public view
	Concerned about the disadvantaged groups, including rural people, the unemployed, and the disabled	EXP21 Concerned about the disadvantaged groups
	The media treats the readers/audiences in an equal manner, instead of being arrogant	EXP22 Equal manner

(continued)

Table 9.3 (continued)

Categories	Statement in questionnaire	Symbols in data analysis
	The news reports tell the truth, not using high-flown rhetoric	EXP23 Not high-flown
	Not reporting some conferences and government affairs that have little to with the ordinary people	EXP24 Reducing conference news
Media ethics	The proportion of advertising is appropriate	EXP25 Controlling proportion of advertising
	Not posting false advertising	EXP26 No false advertising
	Not publishing or broadcasting advertised or paid news, soft-sell advertising	EXP27 No paid news
	The reporting style is elegant, without vulgar news	EXP28 Elegant style
	The editorial is independent from marketing so reports are not affected by money	EXP29 Independent editing
Relationship with the government	Not reporting negative news that may cause social unrest	EXP15 Controlling negative news
	The media is consistent with the Party and the government under all circumstances	EXP16 Consistency
	The media of high authority represent the high-level Party and government offices	EXP17 High authority
	Able to independently review and supervise the government policies and actions	EXP14 Government supervision
External image	The media has a long history	EXP30 Long history
	The media has a large scale circulation and volumes	EXP31 Large scale
	The media are well received by many people	EXP32 Positive review

9.4.1 First Round of Selection Based on Importance Score

9.4.1.1 Explanation of the Standard Referenced in the First Round of Selection

(1) *Mean of importance*

According to the statement, 1-point means "Very Unimportant", 2 points "Unimportant", 3 points "Slightly Important", 4 points "Fairly Important", and 5 points "Very

Important". Then the scores of 32 items given by respondents represents the importance of each item. 3 points can be regarded as the dividing line of importance. In addition, the average score of 32 items is 3.87, which can be referenced as a standard. If we want to sift through data in a stricter way, the standard should be improved. However, the importance score is not exclusively one of the criteria, and other sectors would be considered in the analysis.

(2) *Factor loading*

By performing a factor analysis that extracts only one factor from the importance evaluation results (using the principal components), the factor loading of each item can be obtained. In statistics, the factor loading represents each observation variable and common factor. The greater the absolute value of the factor loading, the closer the relationship between the variable and the common factor. When it comes to this scale, the relationship is the degree of connection between each item and the meaning of the overall scale. Items with low factor loading are probably not used by the public in evaluating the credibility.

(3) *Corrected Item-Total Correlation*

It represents the correlation coefficient of each item and the total score of the scale. In this scale, it has a similar meaning to the factor loading. If the correlation coefficient score is not high, it may not be the standard in evaluating media credibility.

(4) *Alpha if Item Deleted*

The reliability coefficient can be used to analyse the internal consistency of each item. The most commonly used "Cronbach's α" is adopted. "The Alpha if Item Deleted" can indicate the change of the coefficient α after this item is deleted. If the α increases significantly, this item has a large negative impact on the internal consistency of the scale. That is to say, the item is not consistent with the scale so that such item is not be the basis for judging the media credibility.

9.4.1.2 The Criteria for the First Round of Selection

The primary standard for classification is the mean value of importance scores, and scale consistency data is used as a reference. The item is selected from two perspectives: Is it a criterion for media credibility? Is it an important criterion for media credibility?

In this way the selection of items can be divided into three categories:

Type① (non-criteria): not a criterion for media credibility (of course, it is not an important criterion).

The selection criterion of statistical data: the average score is lower than 3 points; the consistency with the scale is very poor (factor loading and correlation coefficient are lower than 0.3 and α is higher than 0.8845). It is represented by the symbol "①", referred to as "non-standard".

Type② (non-important criteria): Criteria that can be used to evaluate credibility, but not important ones.

The selection criterion of statistical data: mean value is higher than 3, but the consistency with the overall scale is not good (the factor loading is lower than 0.4855; correlation coefficient is lower than 0.4377; α is higher than or close to 0.885). It is represented by the symbol "②", referred to as "non-important criteria".

Type③ (important criteria): an important criterion for credibility evaluation.

The selection criterion of statistical data: the mean score is close to 4 or higher, and consistent with the scale (the factor loading is higher than 0.485, the correlation coefficient is higher than 0.4377, and the α is lower than 0.885). Use the symbol "③", referred to as "important standard".

In addition, some items only partially meet the statistical standard, and they cannot be clearly classified into a certain category. If they are closer to the high-authority category, they are represented by "+", and if closer to lower level, represented by "-". The exact attribution will be discussed in later analysis.

It should be emphasized that statistical data is only a reference. Although different data provide evidences from different angles, it is by no means a rigid frame or golden rule. The conclusion only means that "the item has more possibilities and supports", but it does not mean that it is a general truth.

9.4.1.3 The Results of the First Round of Selection

The statistics results of the 32 items are shown in Table 9.4 (the items are arranged in descending order of the mean importance score).

9.4.1.4 Analysis and Interpretation of the First Round of Selection Results

(1) Items that are not suitable for the evaluation criteria (*Type① "non-criteria"*).

Three items have a mean score lower than 3, and their factor loading and correlation are quite low. Therefore, the last three questions in Table 9.4: "EXP30 The media has a long history (mean 2.82)"; "EXP16 The media is consistent with the Party and the government under all circumstances (mean 2.59)"; "EXP17 The media of high authority represents the higher-level Party and government offices (mean 2.45)" can be regarded as unsuitable in evaluating media credibility, and classified as "Type①".

In addition, the two items, "EXP15 Not reporting negative news that may cause social unrest (mean 3.14)" and "EXP24 Not reporting some conferences and government affairs that have little to with the ordinary people (mean 3.43)", have a low score in factor loading and correlation which means a low consistency with the topic "whether the media is credible", although their mean values are slightly higher than 3. Therefore, they are not considered as the standard of evaluation, and classified as "Type①".

Table 9.4 The results of the first round of selection (sample size = 689)

Items	Mean	Factor loading	Corrected item-total correlation	Alpha if item deleted (α)	Results of first selection
EXP01 Factual report	4.49	0.564	0.440	0.884	③
EXP20 Public view	4.43	0.635	0.536	0.883	③
EXP26 No false advertising	4.39	0.501	0.410	0.885	③–
EXP22 Equal manner	4.37	0.674	0.556	0.882	③
EXP02 Full coverage	4.36	0.524	0.409	0.885	③–
EXP23 No high-flown	4.32	0.631	0.497	0.883	③
EXP11 Correcting errors	4.31	0.628	0.554	0.883	③
EXP18 Respect for privacy	4.30	0.547	0.459	0.884	③
EXP21 Concerned about the disadvantaged groups	4.28	0.577	0.487	0.884	③
EXP04 Accurate report	4.26	0.597	0.501	0.883	③
EXP07 Prompt and timely	4.26	0.601	0.521	0.883	③
EXP06 Balancing without bias	4.26	0.624	0.520	0.883	③
EXP14 Government supervision	4.20	0.491	0.383	0.885	③–
EXP05 Objective without bias	4.13	0.556	0.471	0.884	③
EXP12 Focus on reality	4.07	0.591	0.532	0.883	③
EXP13 Sharp perspective	4.06	0.635	0.572	0.882	③
EXP19 Exposing wrongdoing	4.04	0.520	0.441	0.884	③
EXP28 Elegant style	4.03	0.563	0.508	0.883	③
EXP29 Independent editing	3.96	0.550	0.473	0.883	②+
EXP03 Complete coverage	3.91	0.574	0.495	0.883	②+
EXP25 Controlling proportion of advertising	3.83	0.414	0.359	0.886	②
EXP08 In-depth	3.82	0.538	0.486	0.883	②
EXP32 Positive review	3.70	0.376	0.382	0.885	②
EXP27 No paid news	3.61	0.468	0.439	0.884	②
EXP10 Novel and vivid	3.50	0.432	0.451	0.884	②
EXP24 Reducing conference news	3.43	0.312	0.286	0.888	①
EXP31 Large scale	3.38	0.354	0.396	0.885	②–
EXP09 Exclusive	3.19	0.367	0.389	0.885	②–
EXP15 Controlling negative news	3.14	0.184	0.259	0.889	①
EXP30 Long history	2.82	0.275	0.338	0.887	①
EXP16 Consistency	2.59	0.154	0.260	0.889	①
EXP17 High authority	2.45	0.080	0.195	0.891	①
Mean of 32 items	3.8716	0.4855	0.4377	0.885	

Notes In the results ① indicates that the item is not suitable for evaluating credibility. ② Indicates could be criteria but not an important one. ③ indicates represents an important standard for credibility. Items between the two categories are marked with "+" and "–", and their classification needs further evidence

(2) Items that can be used as criteria are not important criteria (*Type*② "*non-important criteria*").

"EXP25 The proportion of advertising is appropriate (mean 3.83)"; "EXP08 Able to provide in-depth analysis and interpretation of complex news events (mean 3.82)"; "EXP32 The media are well received by many people (mean 3.70)", "EXP27 Not publishing or broadcasting advertised or paid news, soft-sell advertising (mean 3.61)"; "EXP10 The news layout or program is novel, vivid and unique (mean 3.50)", "EXP31 The media has a large scale the circulation and volumes (mean 3.38) " and " EXP09 Able to have exclusive news often (mean 3.19) ". The mean value of the 7 items is lower than the overall average 3.8716, but their correlation scores are acceptable. They could be used in determining the credibility, but are not important. They are classified as "Type②".

If the higher standard is adopted, the two items with an average score of lower than "4 fairly important" would also be included in "non-important criteria", which are "EXP 29 The editorial is independent from marketing so reports are not affected by money (mean 3.96) "and "EXP03 Report all news events that the public wants to know (mean 3.91) ". They are classified as "Type②+".

(3) Important criteria in evaluation (*Type*③ "important criteria").

Among the remaining 18 items, three items, "EXP14 Able to independently review and supervise the government policies and actions"; "EXP02 Full coverage of news events, not avoiding any important facts" and "EXP26 Not posting false advertising", have a relatively poor performance in correlation and classified as "Type③-". The rest of them are well performed in mean score and correlation so that they are classified into the "Type③".

9.4.2 Second Round of Selection Based on Distribution of Score

The first round of selection mainly relies on the mean value of each item. In descriptive statistics, mean as a very important statistic reflects the overall evaluation, but cannot describe the specific distribution of each observation. In some cases, especially when the observations are skewed, the mean may hide some important facts. For example, in this scale, when the average value is 3 points, the distribution of opinions may be very different. One is that many respondents prefer 3; the other is that half of respondents score 1 and the other half score 5. Obviously, the two cases should not be treated equally. It is necessary to further select and analyse the evaluation criteria based on the distribution of each item.

9.4.2.1 Explanation of the Standard Referenced in the Second Round of Selection

The objective of developing the "Importance Scale" is to obtain the criteria that are widely recognized as "important", that is, the more concentrated to the "4 fairly important" and "5 very important", which are the items selected to evaluate media credibility. Skewness in descriptive statistics is used to reflect the skewness of the distribution, so the main reference statistic for the second round of selection is skewness. In order to have a clear and direct observation and analysis, bar graphs are adopted.

(1) *Mean*

It is used to reflect the overall distribution of the score.

(2) *Skewness*

The skewness is used to indicate the degree of skewness of observation. The 0 skewness indicates a normal distribution. Smaller than zero means negatively skewed, and bigger is positively skewed. For the "Importance Scale, it is used to represent the distribution of scores on the scale of 1–5.

The larger the absolute value of skewness, the more concentrated the respondents' opinions are. Specific to this scale, the observation is 1–5. The higher the absolute value of skewness, the more concentrated the respondent's opinion, i.e. the higher the consistency of opinions. The smaller the absolute value, the more dispersed the opinions are, i.e. the more divergent the opinions are.

A negative skewness indicates that the respondents prefers a high score (5 very important). A positive skewness is the opposite (a score of 3 or less). Therefore, the smaller the skewness value, the important the criterion is.

Bar graph is used to reflect the distribution in a direct way. The results of first round of selection are listed in the results, which can be referred for a comprehensive analysis.

9.4.2.2 The Criteria for the Second Round of Selection

The selection of each item can be interpreted from three perspectives. Is it a criterion for evaluating media credibility? Is it an important criterion for media credibility evaluation? Is it a widely accepted criterion for media credibility evaluation?

From the concentration of observation, the items can be divided into four categories.

Type I: respondents' opinions are clearly concentrated in "1 very unimportant". It shows most people take the item as not important in evaluation, so the item is not suitable as a criterion for. Such items are represented by the symbol "I".

Type II: opinions are not concentrated. Results are distributed in a balanced manner between the two ends. It appears a tit-for-tat situation. Towards one behaviour of the media, a number of people regards it as trustworthy, but the other hold completely opposite opinions. Such an item is not suitable as a criterion for evaluating the credibility of the overall media. However, the completely opposed opinion may reflect a particular case of the China's media at this stage. Such items need to be further analysed. It is represented by "II".

Type III: no opposition appears. The importance of the item is recognised, but its degree is varied. Results are evenly spread among "slightly important", "fairly important" and "very important. It means that the public regards it as a criterion, but it is not widely accepted that it is "very important". Such item is represented by the symbol "III".

Type IV: results are obviously concentrated at the end of "5 very important", so the people agree with this item is an important criterion for credibility. Use the symbol "IV" to represent.

For some items that cannot be separated, they are marked with the same symbols in first round of selection, "+" and "−".

9.4.2.3 Results of the Second Round of Selection

The specific analysis results are shown in Table 9.5 (the order of the items is arranged in descending order of the overall average).

9.4.2.4 Analysis and Interpretation of the Results of the Second Round of Selection

The situation according to the distribution of results can be divided into the following types.

(1) *Concentrated on the end of "not important" (Type I)*

A positive skewness of the item indicates that most of the respondents' opinions are towards "1 very unimportant". There are two obvious positive outliers:

EXP17 The media of high authority represent the higher-level Party and government offices.

EXP16 The media is consistent with the Party and the government under all circumstances.

Most believe that two items are unimportant in evaluating the credibility, so they are not suitable as criteria. They are marked as Type I, which matches the results of first round of selection, "non-standard". The distributions of opinions on five different degrees are shown in Figs. 9.1 and 9.2.

Overall, about 20% of samples consider the media's "high authority" and "consistency with the government" as a "fairly important and "very important" standard in evaluating its credibility. It validates the interview results before the survey. Opinions like "this newspaper is published by the Party's Central Office, so it is credible" and "the newspaper is trustworthy as it is run by the government" appeared in former interviews.

Table 9.5 The results of the second round of selection

Item	Overall sample		Selection results	
	Mean	Skewness	First	Second
EXP01 Factual report	4.49	-1.83	③	IV +
EXP20 Public view	4.43	-1.58	③	IV +
EXP26 No false advertising	4.39	-1.49	③–	IV +
EXP22 Equal manner	4.37	-1.41	③	IV +
EXP02 Full coverage	4.36	-1.45	③–	IV +
EXP23 Not high-flown	4.32	-1.43	③	IV +
EXP11 Correcting errors	4.31	-0.92	③	IV
EXP18 Respect for privacy	4.30	-1.28	③	IV +
EXP21 Concerned about the advantaged group	4.28	-1.26	③	IV +
EXP04 Accurate report	4.26	-1.14	③	IV +
EXP07 Prompt and timely	4.26	-1.24	③	IV +
EXP06 Balancing without bias	4.26	-1.23	③	IV +
EXP14 Government supervision	4.20	-1.17	③–	IV +
EXP05 Objective without bias	4.13	-0.94	③	IV
EXP12 Focus on reality	4.07	-0.72	③	IV-
EXP13 Sharp perspective	4.06	-0.72	③	IV-
EXP19 Exposing wrongdoing	4.04	-0.91	③	IV
EXP28 Elegant style	4.03	-1.01	③	IV
EXP29 Independent editing	3.96	-0.87	②+	IV-
EXP03 Complete coverage	3.91	-0.74	②+	IV-
EXP25 Controlling proportion of advertising	3.83	-0.64	②+	III
EXP08 In-depth	3.82	-0.61	②	III
EXP32 Positive review	3.70	-0.62	②	III
EXP27 No paid news	3.61	-0.35	②	III-
EXP10 Novel and vivid	3.50	-0.33	②	III-
EXP24 Reducing conference news	3.43	-0.28	①	III-
EXP31 Large scale	3.38	-0.34	②–	III-
EXP09 Exclusive	3.19	0.01	②–	II +
EXP15 Controlling negative news	3.14	-0.09	①	II
EXP30 Long history	2.82	0.19	①	II
EXP16 Consistency	2.59	0.42	①	I
EXP17 High authority	2.45	0.56	①	I

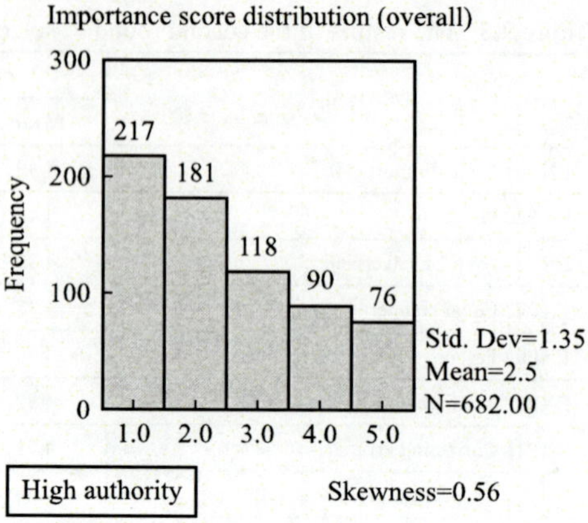

Fig. 9.1 . *Notes* 1.0 = very unimportant; 2.0 = not important; 3.0 = slightly important; 4.0 = fairly important; 5.0 = very important

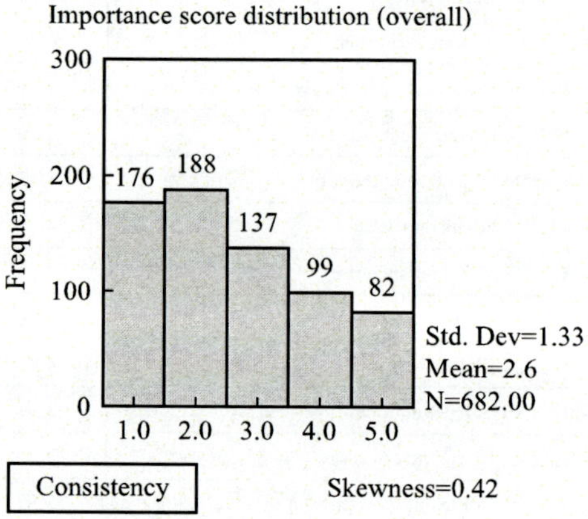

Fig. 9.2 . *Notes* 1.0 = very unimportant; 2.0 = not important; 3.0 = slightly important; 4.0 = fairly important; 5.0 = very important

It reflects changing of people's perceptions of the media during the social transformation and news reforming in China. This special issue is worthy of further research in media credibility. It will be analysed with the support of other data in the following section.

(2) *Opposite-shaped items (Type II)*

In the column of overall skewness, the closer the absolute value of skewness to 0, the more dispersed the opinions are. There are 3 outliers.

EXP15 Not reporting negative news that may cause social unrest.

EXP30 The media has a long history.

EXP09 Able to have exclusive news often.

The absolute value of the skewness of the three items is between 0.01 and 0.19. The following Figs. 9.3, 9.4 and 9.5 show the data more directly.

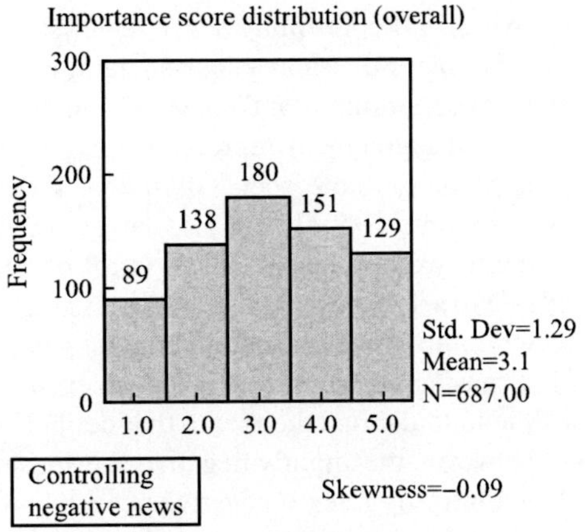

Fig. 9.3 . *Notes* 1.0 = very unimportant; 2.0 = not important; 3.0 = slightly important; 4.0 = fairly important; 5.0 = very important

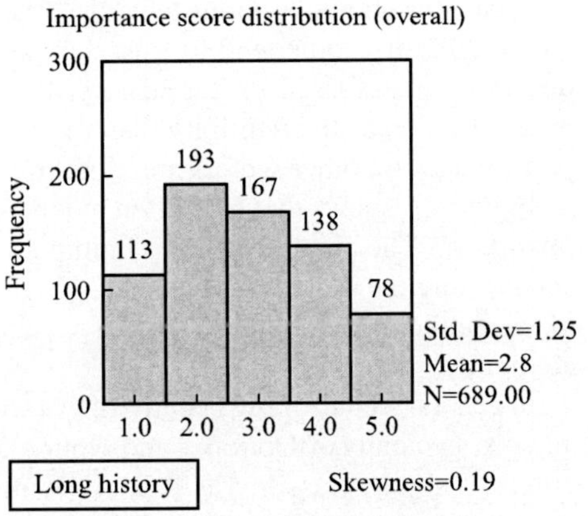

Fig. 9.4 . *Notes* 1.0 = very unimportant; 2.0 = not important; 3.0 = slightly important; 4.0 = fairly important; 5.0 = very important

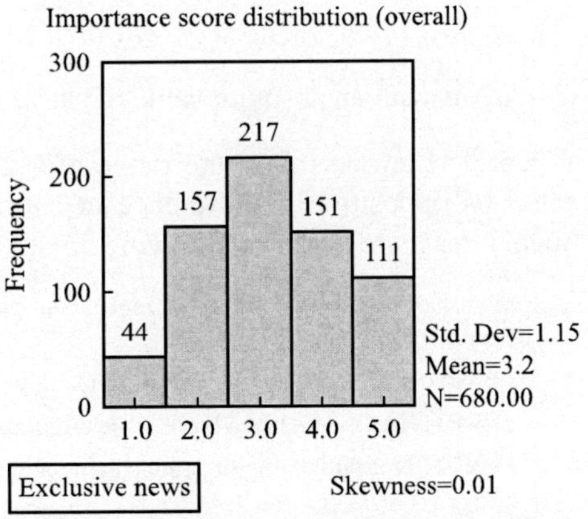

Fig. 9.5 . *Notes* 1.0 = very unimportant; 2.0 = not important; 3.0 = slightly important; 4.0 = fairly important; 5.0 = very important

In Fig. 9.3 "controlling negative news" there is a clear disagreement on controlling negative news that may cause social unrest. Many people agree "5 very important" and "1 very unimportant", and the two ends are almost with the same weight.

This disagreement represents the conflicts on whether negative reports would damage or promote social harmony. Indeed, "not causing social unrest" is used as a reason to blindly suppress negative news reports. Reports during the SARS epidemic exemplified it. The control of negative news is so excessive that reports related to safety of people are suppressed, which inevitably has a negative impact on the credibility of the media. If the respondents realize this situation and thinks that "not causing social unrest" is just an excuse to suppress the freedom of reports, they will tend to distrust the media that deliberately control negative news.

However, the slightly negative skewness approves people don't trust the media if they arbitrarily release negative news regardless of social stability. It seems that the media's social responsibility and constructiveness are important.

Figure 9.4 shows opinions towards "(the media has) a long history" are scattered. About 30% of people tend to trust the media with a long history, but more people do not evaluate based on the history. It reflects the old media does not have too many advantages in credibility than the newly established brands. For the majority, performance on other sectors are more important.

Figure 9.5 shows that for "often having exclusive news", most agreed "3 slightly important". The number of "very unimportant" is lower than that of other two items, "controlling the negative news" and "high authority". It indicates that in the eyes of the public, whether there is often exclusive news is not very important to media credibility.

In general, "controlling negative news and "(with) a long history" shows a controversy at two ends of Option 1 and Option 5, so they are not suitable for evaluation criteria. They are classified as Type II. Opinions on the "often having exclusive news" are not concentrated but balanced, so it is classified as Type II+.

(3) *Neither opposing nor concentrated enough (Type III).*

(1) Items with an absolute value of skewness between 0.28 and 0.35.

When the skewness is in the range of 0.28–0.35, the disagreement still exists. It could be opposition or difference in consistency, which require further analysis. Such 4 items are listed as following.

EXP24 Not reporting some conferences and government affairs that have little to do with the ordinary people.

EXP10 The news layout or program is novel, vivid and unique.

EXP31 The media has a large scale circulation and volumes.

EXP27 Not publishing or broadcasting advertised or paid news, soft-sell advertising.

The distribution graph is adopted in interpretation.

Compared with the opposing at two ends, there are fewer people who hold "1 very unimportant" and "2 not important" attitudes on "large scale" and "novel and vivid form", and the opposition is not obvious. Opinions are more concentrated on "3 slightly important and "4 fairly important" (see Figs. 9.6 and 9.7).

Slightly different from the above, more people regard "not reporting conferences that have little to do with the ordinary people" and "no paid news" as "very important". More people are not concerned about the unimportant conferences and repugnant to paid news, so that they can greatly reduce the trust in the media. Generally speaking, the weight of "unimportant" has begun to be lower than that of "important", so it is difficult to form a tit for tat. Moreover, opinions on "controlling paid news" have concentrated on the "important" end (see Figs. 9.8 and 9.9).

Overall, for the above four items, the opposition is not very obvious. The number of "1 very unimportant" and "2 not important" is lower than the three items analysed in former section ("consistency", "high authority" and "long history"). It means people recognize these factors as important, even though their opinions are not strong and concentrated. There are large differences in the degree of importance. These items

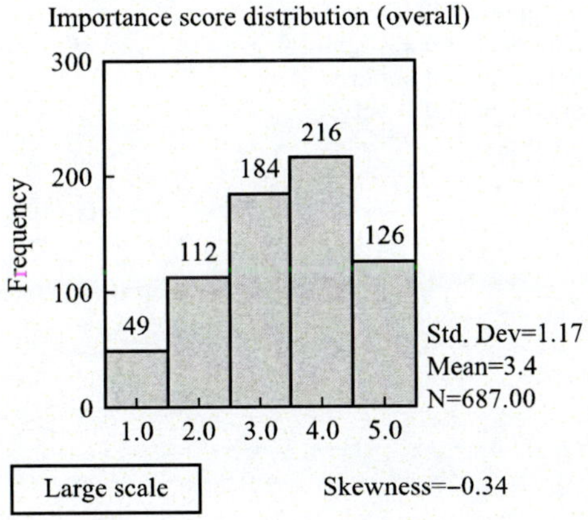

Fig. 9.6 . *Notes* 1.0 = very unimportant; 2.0 = not important; 3.0 = slightly important; 4.0 = fairly important; 5.0 = very important

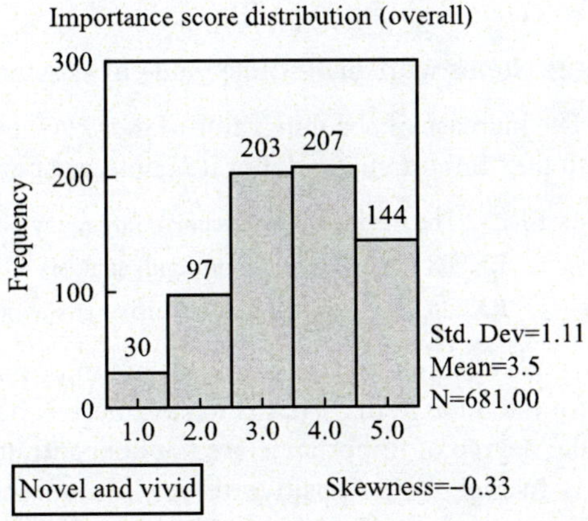

Fig. 9.7 . *Notes* 1.0 = very unimportant; 2.0 = not important; 3.0 = slightly important; 4.0 = fairly important; 5.0 = very important

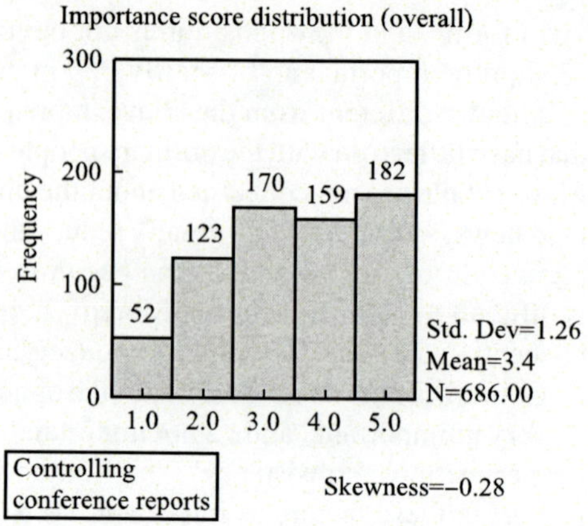

Fig. 9.8 . *Notes* 1.0 = very unimportant; 2.0 = not important; 3.0 = slightly important; 4.0 = fairly important; 5.0 = very important

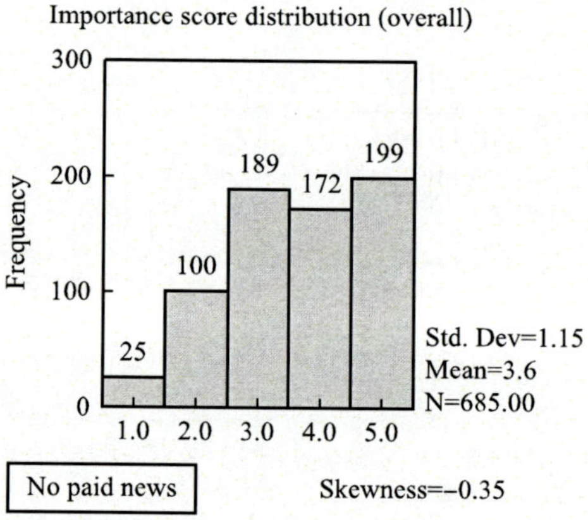

Fig. 9.9 . *Notes* 1.0 = very unimportant; 2.0 = not important; 3.0 = slightly important; 4.0 = fairly important; 5.0 = very important

can be deemed as slightly opposed, less important, and less concentrated opinions, and classified into Type III-.

(2) Items with an absolute value of skewness between 0.61 and 0.64.

The increase of absolute value of the skewness means that the opinion is more skewed to the "important" end, but it is not very concentrated. Here are three such items:

EXP32 The media are well received by many people.

EXP08 Able to provide in-depth analysis and interpretation of complex news events.

EXP25 The proportion of advertising is appropriate.

We can find that the number "not important" is far lower than that of "important" for the three items. That is to say, there is basically no oppositions, and opinions on the degree of importance are not concentrated enough, distributed on 3, 4, and 5.

In Fig. 9.10, "positive review" represents the influence of the opinions of the surrounded people on the evaluation of media credibility. It seems "good reputation" has a more positive effect on the credibility.

Fig. 9.10 . *Notes* 1.0 = very unimportant; 2.0 = not important; 3.0 = slightly important; 4.0 = fairly important; 5.0 = very important

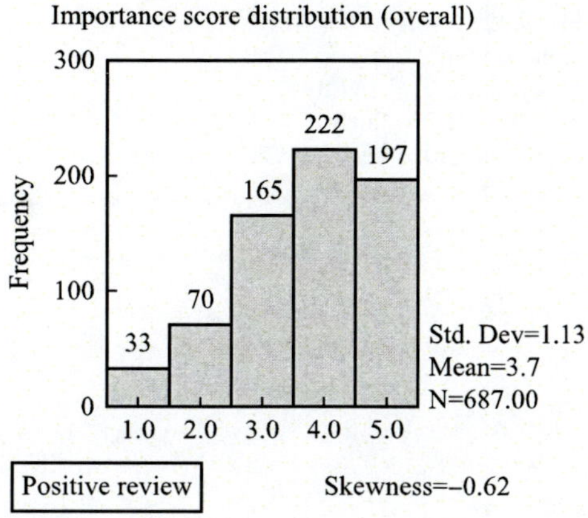

In Figs. 9.11 and 9.12, the number of "fairly important" and "very important" have increased on "the frequency of in-depth reports" and "controlling the proportion of advertising", even though the results are still not very concentrated. In former interviews, some respondents expressed the advertisements are understandable, like "the media relies on advertising to survive." Based on the overall data, if the advertising is too flooded, it will still have a negative impact on media credibility.

The above three items can be classified into Type III. People agree that they are important, but they are no consensus on its high importance.

(4) More concentrated on the importance (Type IV)

When the absolute value of the skewness is above 0.7, it is considered that it has obvious skewness. If the absolute value is greater than 1, the skewness is very serious. For the "Importance Scale", a seriously negatively skewed item is a widely accepted important criterion for media credibility.

① Items with an absolute value of skewness between 0.7 and 0.9.

Fig. 9.11 . *Notes* 1.0 = very unimportant; 2.0 = not important; 3.0 = slightly important; 4.0 = fairly important; 5.0 = very important

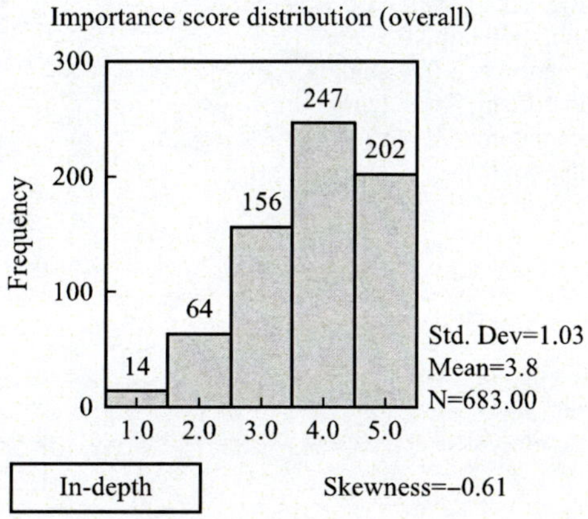

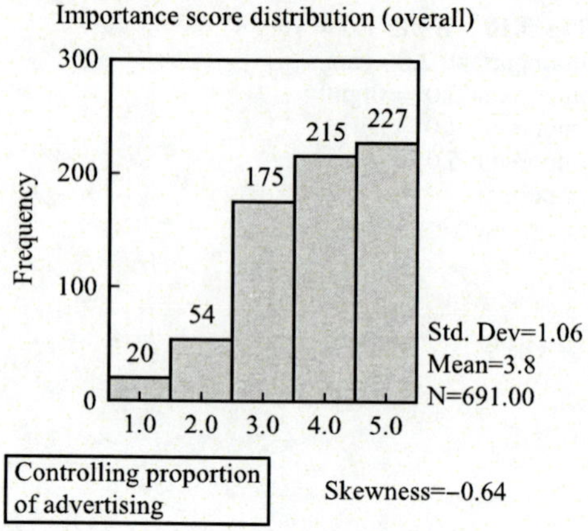

Fig. 9.12 . *Notes* 1.0 = very unimportant; 2.0 = not important; 3.0 = slightly important; 4.0 = fairly important; 5.0 = very important

There are 4 items with an absolute value of skewness between 0.7 and 0.9:

EXP03 Report all news events that the public wants to know.

EXP12 The news covers the most realistic and urgent issues in social development.

EXP13 Good at finding problems and having a sharp perspective.

EXP29 The editorial is independent from marketing so reports are not affected by money.

There are fewer "not important" results, and the majority is concentrated on "4 fairly important" and "5 very important". The "3 slightly important" is also chosen by many people. It is a widely accepted criterion for judging the credibility of the media. Since the opinions are not concentrated enough, these four items are classified into the Type IV- (see Figs. 9.13, 9.14, 9.15 and 9.16).

② Items with an absolute value of skewness of about 1.

There are 4 items' absolute value of skewness close to 1.

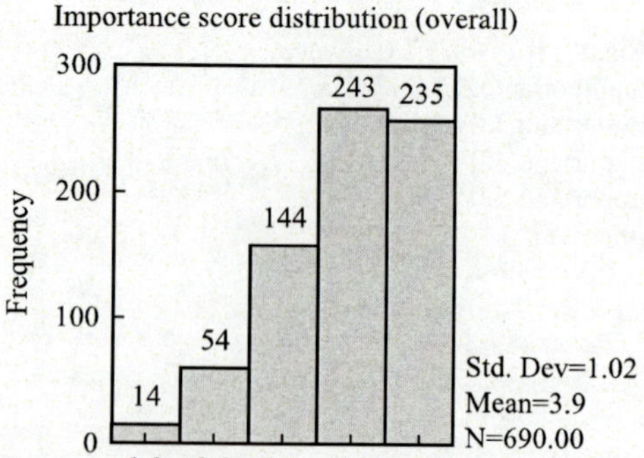

Fig. 9.13 . *Notes* 1.0 = very unimportant; 2.0 = not important; 3.0 = slightly important; 4.0 = fairly important; 5.0 = very important

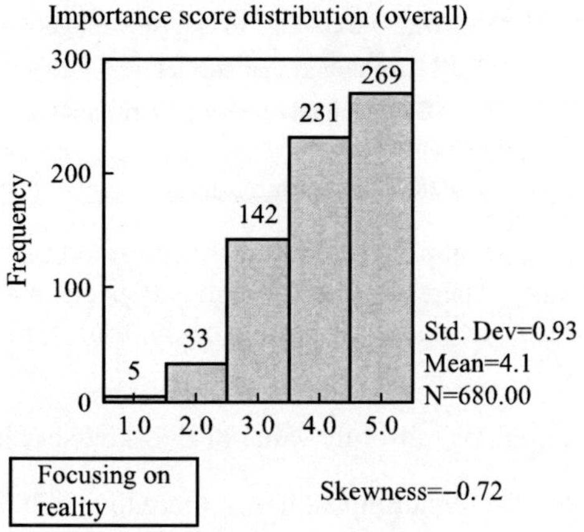

Fig. 9.14 .
Notes 1.0 = very unimportant; 2.0 = not important; 3.0 = slightly important; 4.0 = fairly important; 5.0 = very important

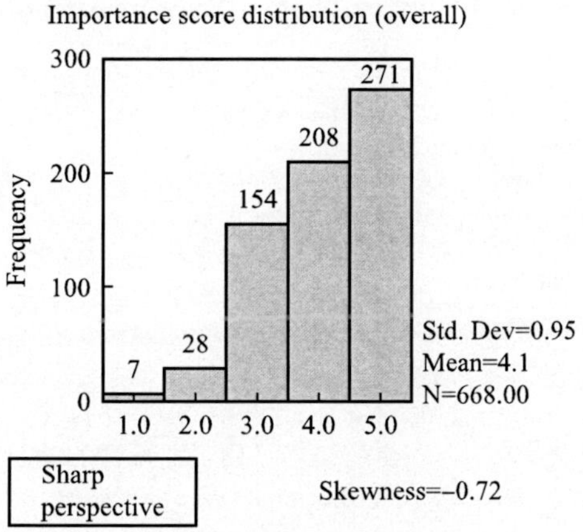

Fig. 9.15 .
Notes 1.0 = very unimportant; 2.0 = not important; 3.0 = slightly important; 4.0 = fairly important; 5.0 = very important

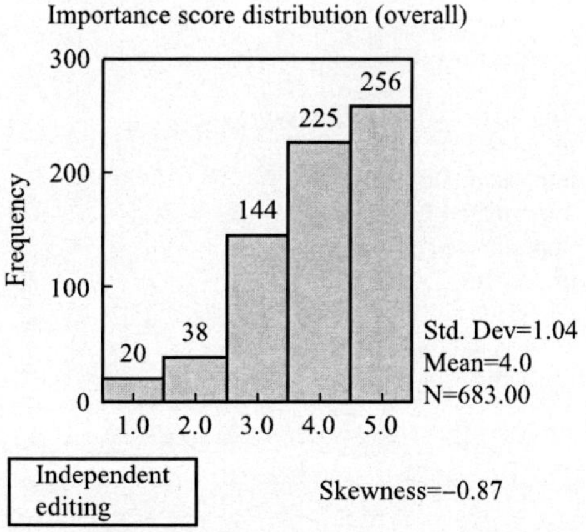

Fig. 9.16 .
Notes 1.0 = very unimportant; 2.0 = not important; 3.0 = slightly important; 4.0 = fairly important; 5.0 = very important

EXP19 With the courage to expose wrongdoing and high proportion of critical reports.

EXP11 Apologise and correct wrong reports once they have been found.

EXP05 Objectively present the original appearance of the event without adding the bias of the reporter.

EXP28 The reporting style is elegant, without vulgar news.

The opinions are more concentrated on "4 slightly important" and "5 very important". There is still no obvious concentration on 5, so they are classified as IV. The distributions of the results of the four items are shown in Figs. 9.17, 9.18, 9.19 and 9.20.

③ Items' absolute value of the skewness is significantly greater than 1.

For the remaining 12 items, more than half of the opinion falls on "5 very important". The concentration trend is obvious and can be considered as the most accepted criteria for credibility evaluation. They are categorised in Type IV+ (see Table 9.6).

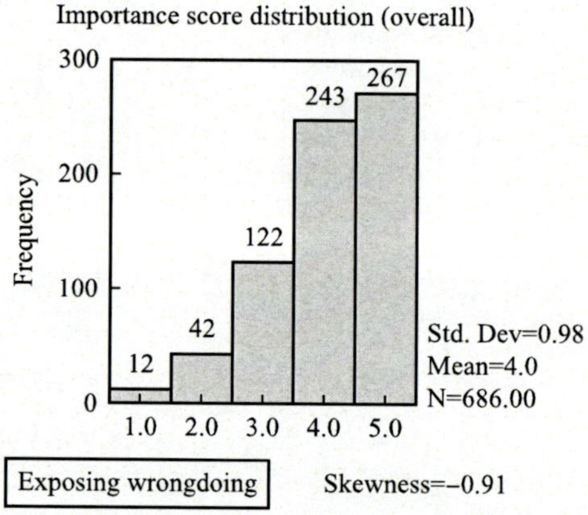

Fig. 9.17 . *Notes* 1.0 = very unimportant; 2.0 = not important; 3.0 = slightly important; 4.0 = fairly important; 5.0 = very important

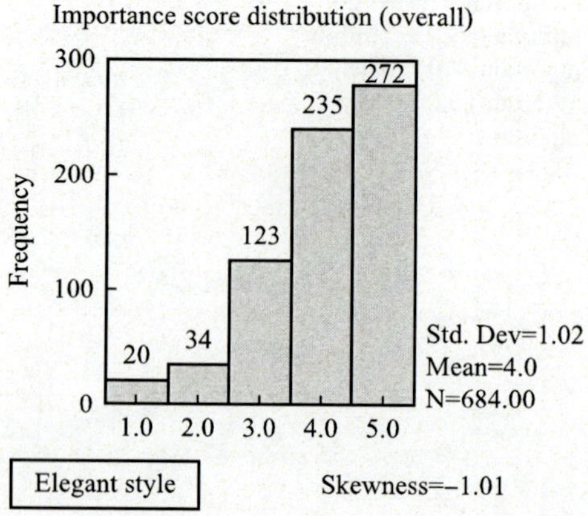

Fig. 9.18 . *Notes* 1.0 = very unimportant; 2.0 = not important; 3.0 = slightly important; 4.0 = fairly important; 5.0 = very important

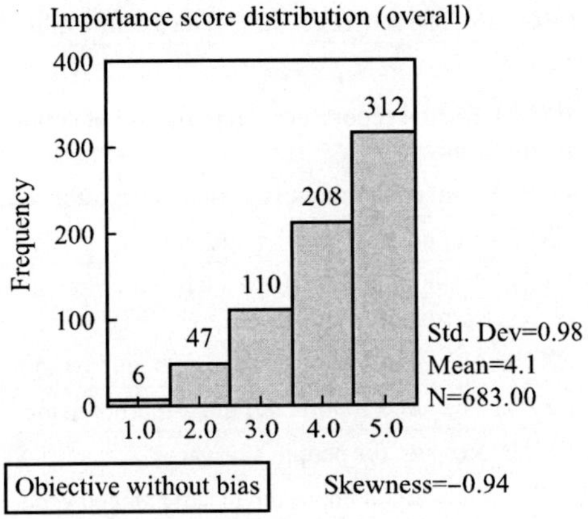

Fig. 9.19 . *Notes* 1.0 = very unimportant; 2.0 = not important; 3.0 = slightly important; 4.0 = fairly important; 5.0 = very important

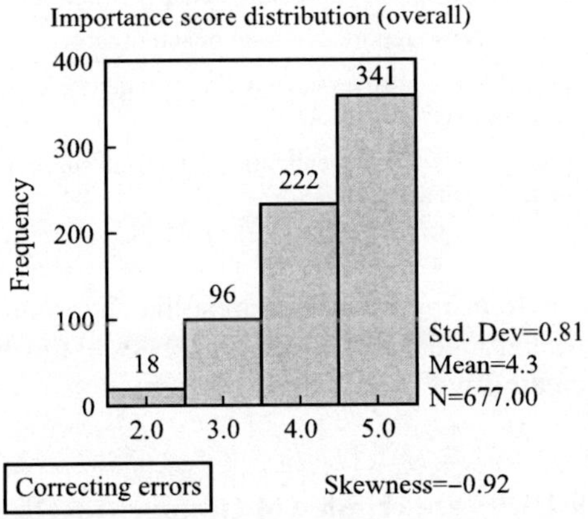

Fig. 9.20 . *Notes* 1.0 = very unimportant; 2.0 = not important; 3.0 = slightly important; 4.0 = fairly important; 5.0 = very important

9.4.3 Correction of Results and Analysis of Their Characteristics Based on the Differences Among Education Backgrounds

If the data in the previous two rounds of selections come from random sample, the conclusions can be used as a preliminary basis for evaluating the credibility of China's media. However, the data is non-random sampling from Beijing. Half of the samples are students at Tsinghua University and Renmin University of China. The proportions of highly educated and student are obviously higher than that from the whole country. Many studies have proved that people with different educational backgrounds have different opinions towards the criteria for evaluating media credibility. Therefore, the results obtained in this research will be different from the overall situation of China. It is necessary to further analyse the criteria according to their education backgrounds, and correct the deviation caused by the particularity of the sample.

Table 9.6 The most accepted items in the evaluation of credibility

Item	Mean	Skewness
EXP01 Factual report, news reports do not contain false, speculative and fictitious elements	4.49	−1.83
EXP20 Concerned about the public interest in their view	4.43	−1.58
EXP26 Not posting false advertising	4.39	−1.49
EXP22 The media treat the readers/audiences in an equal manner, instead of being arrogant	4.37	−1.41
EXP02 Full coverage of news events, not avoiding any important facts	4.36	−1.45
EXP23 The news reports tell the truth, not using high-flown rhetoric	4.32	−1.43
EXP18 Respect for people's privacy	4.30	−1.28
EXP21 Concerned about the disadvantaged groups, including peasants, the unemployed, and the disabled	4.28	−1.26
EXP04 The news report is particularly accurate	4.26	−1.14
EXP07 News reports are prompt and timely	4.26	−1.24
EXP06 For controversial news, balancing the coverage of each party without bias	4.26	−1.23
EXP14 Able to independently review and supervise the government policies and actions	4.20	−1.17

More importantly, comparing the opinions of groups with different education backgrounds is an in-depth analysis of the difference of criteria for evaluating media credibility.

9.4.3.1 Description of Groups with Different Education Backgrounds

According to the previous research, education background has a clear correlation with the evaluation of media credibility. In addition, for groups who have received college education, there is great difference in their opinions about media credibility among undergraduates, graduates and the ones taking a job after graduation. This study divides the 692 samples into three groups based on education backgrounds according to the education level and life experience. The detail of groups is explained as followings.

(1) *Group of "undergraduates".*

They are mainly undergraduate students under 22 years old at the Renmin University of China and Tsinghua University. Among them 16–17 years old account for 3.5%, 18 years old 31.1%, 19 years old 25.9%, 20 years old 23.3%, and 21 years old 16.2%. In terms of education background, they have not yet obtained a bachelor's degree; from the social experience, most of them have just entered into the society, and are accumulating the social experience.

(2) *Group "lower than a bachelor's degree".*

Table 9.7 The age distribution of groups with different educational backgrounds

Group	Frequency	Proportion (%)	Composition of ages and its proportion
Undergraduates	228	32.9	16–17 years old, 3.5% 18 years old, 31.1% 19 years old, 25.9% 20 years old, 23.3% 21 years old, 16.2%
Lower than a bachelor's degree	217	31.3	16–17 years old, 1.8% 18–35 years old, 76.5% 36–50 years old, 14.7% Over 50 years old, 6.9%
With a bachelor's degree or higher	247	35.8	Lower than35 years old, 90.7% 36–50 years old, 7.3% Over 50 years old 2.0%
Total	692	100%	

It refers to people over 16 years having education lower than a bachelor's degree. It includes four high school students aged at 16–17 years, accounting for 1.8%; 18–35 years old 76.5%, 36–50 years old 14.7%, and people over 50 years old 6.9%.

(3) *Group with "a bachelor's degree or higher".*

For people aged at 22 and above with a bachelor's degree or higher. It mainly includes senior undergraduates, graduate students and ones who have a job. Among them, people under 36 years old account for 90.7%, 36–50 years old 7.3%, and over 50 years old 2.0% (see Table 9.7).

The composition of ages reflects that the subjects is at the middle age. Also they are the most influential people in society.

For the actual composition of Chinese population, people lower than a bachelor's degree is the majority, which has quantitative advantages. People with a bachelor's degree or higher is better at expressing opinions and acting in society, and have the advantage of the influence of opinion.

9.4.3.2 The Overall Comparison of the Criteria for Evaluating Media Credibility Among Different Groups

The mean and skewness of the criteria of the three groups for media credibility are compared, and a one-way analysis of variance is applied. The comparison results of are shown in Table 9.8.

At the 95% confidence interval, the mean of 15 items is not significantly different, which means that groups with different education backgrounds have similar opinions towards them. Their analysis results of the samples can be adopted. In addition, the

Table 9.8 Scores given by groups with different education backgrounds

Item	Undergraduates		Lower than bachelor's degree		With a bachelor's degree or higher		Overall			Results of selection	
	Mean	Skewness	Mean	Skewness	Mean	Skewness	Mean	Skewness	P-value	First round of selection	Second round of selection
EXP01 Factual report	4.55	-1.85	4.46	-1.45	4.66	-2.00	4.49	-1.83	0.000	③	IV +
EXP20 Public view	4.45	-1.51	4.48	-1.51	4.53	-1.67	4.43	-1.58	0.147	③	IV +
EXP26 No false advertising	4.37	-1.12	4.44	-1.31	4.56	-1.99	4.39	-1.49	0.001	③–	I V +
EXP22 Equal manner	4.45	-1.22	4.38	-1.39	4.43	-1.24	4.37	-1.41	0.056	③	I V +
EXP02 Full coverage	4.36	-1.44	4.34	-1.33	4.52	-1.29	4.36	-1.45	0.002	③–	IV +
EXP23 Not high-flown	4.36	-1.35	4.31	-1.23	4.49	-1.41	4.32	-1.43	0.002	③	IV +
EXP11 Correcting errors	4.33	-0.85	4.36	-0.94	4.35	-0.95	4.31	-0.92	0.677	③	IV
EXP18 Respect for privacy	4.42	-1.34	4.28	-1.04	4.37	-1.44	4.30	-1.28	0.037	③	IV +
EXP21 Concerned about the disadvantaged groups	4.34	-1.29	4.38	-1.26	4.32	-1.07	4.28	-1.26	0.510	③	IV +
EXP04 Accurate report	4.20	-0.79	4.42	-1.06	4.35	-1.25	4.26	-1.14	0.089	③	IV +
EXP07 Prompt and timely	4.33	-1.36	4.25	-0.97	4.39	-1.35	4.26	-1.24	0.004	③	IV +
EXP06 Balancing without bias	4.22	-1.10	4.33	-1.09	4.39	-1.51	4.26	-1.23	0.079	③	IV +
EXP14 Government supervision	4.23	-0.93	4.12	-1.08	4.37	-1.41	4.20	-1.17	0.000	③–	IV +
EXP05 Objective without bias	4.15	-0.73	4.07	-0.76	4.30	-1.29	4.13	-0.94	0.006	③	IV
EXP12 Focus on reality	4.11	-0.81	4.07	-0.58	4.20	-0.61	4.07	-0.72	0.015	③	IV-
EXP13 Sharp perspective	4.12	-0.99	4.05	-0.47	4.16	-0.54	4.06	-0.72	0.101	③	IV-

(continued)

Table 9.8 (continued)

Item	Undergraduates		Lower than bachelor's degree		With a bachelor's degree or higher		Overall			Results of selection	
	Mean	Skewness	Mean	Skewness	Mean	Skewness	Mean	Skewness	P-value	First round of selection	Second round of selection
EXP19 Exposing wrongdoing	4.10	-0.75	4.01	-0.76	4.18	-0.98	4.04	-0.91	0.005	③	IV
EXP28 Elegant style	4.09	-0.75	3.95	-0.80	4.15	-1.30	4.03	-1.01	0.008	③	IV
EXP29 Independent editing	3.99	-0.78	4.01	-0.82	4.08	-0.89	3.96	-0.87	0.073	②+	IV-
EXP03 Complete coverage	3.98	-0.73	4.09	-0.80	3.85	-0.67	3.91	-0.74	0.448	②+	IV -
EXP25 Proportion of advertising	4.00	-0.59	3.63	-0.42	3.97	-0.72	3.83	-0.64	0.000	② +	III
EXP08 In-depth	3.91	-0.66	3.80	-0.28	3.96	-0.89	3.82	-0.61	0.009	②	III
EXP32 Positive review	3.61	-0.54	3.80	-0.67	3.74	-0.68	3.70	-0.62	0.294	②	III
EXP27 Paid news	3.54	-0.26	3.63	-0.23	3.74	-0.55	3.61	-0.35	0.112	②	III-
EXP10 Novel and vivid	3.43	-0.27	3.37	-0.22	3.62	-0.38	3.50	-0.33	0.012	②	III-
EXP24 Conference news	3.33	-0.14	3.42	-0.32	3.58	-0.39	3.43	-0.28	0.253	① +	III-
EXP31 Large scale	3.37	-0.45	3.49	-0.28	3.29	-0.31	3.38	-0.34	0.410	②–	III-
EXP09 Exclusive	3.07	0.25	3.21	0.00	3.30	-0.17	3.19	-0.01	0.066	②–	II +
EXP15 Negative news	3.00	0.15	3.42	0.40	2.97	-0.07	3.14	-0.09	0.000	① +	II
EXP30 Long history	2.68	0.28	2.91	0.11	2.83	0.14	2.82	0.19	0.094	①	II
EXP16 Consistency	2.41	0.60	2.97	0.08	2.43	0.56	2.59	0.42	0.000	①	I
EXP17 High authority	2.29	0.75	2.71	0.24	2.28	0.66	2.45	0.56	0.000	①	I

difference of mean value of the other 17 items is significant, which means the results may be caused by education backgrounds, so it requires further analysis.

9.4.3.3 Characteristics of the Evaluation of Media Credibility Among Different Groups

Table 9.9 shows the 17 items with mean reaching the significance level.

The data listed in Table 9.9 cannot tell the difference between the three groups, so further studies of the differences among groups is needed (see Table 9.10).

From Table 9.10, we can summarize characteristics of criteria for evaluating media credibility.

(1) *Whether they have received higher education affect people's perception of the media.*

There are the significant differences between the group "Lower than a bachelor's degree" and the other two. For the "undergraduates" and the "with a bachelor's degree or higher ", there is a significant difference, except for the item, "EXP26 no false advertising". It indicates that the higher education affects people's perception and understanding of the media.

(2) *Two different concepts of the media coexist.*

Among the 17 items, there is a certain pattern between groups—for the first 14 items, the scores of the group "lower than a bachelor's degree" are lower than the other two groups, while the scores for the last three items are higher than that of them. When the contents of last three items and the first 14 items are compared, we find the difference is mainly from the concept of the media. This can be divided into two different situations.

① Differences in standards caused by opposing media concepts.

Let's take a look at the content of the last three items:

EXP17 The media of high authority represent the higher-level Party and government offices.

EXP16 The media is consistent with the Party and the government under all circumstances.

EXP15 Not reporting negative news that may cause social unrest.

These three items have traces of traditional news value standards, which is related to the special media system. In China the media system is a pyramid-shaped and has a strict hierarchy. The media have certain level of authority. They are responsible for the Communist Party at the same level, and the lower Party Committee is responsible for the higher one. At that time with continuous political movements, political news is the most important information which would change people's life. The role of the media is to publicize the policies and instructions from government. The function of the media is like the mouthpiece of the government. In this social context the media with higher level has the advantage of political information resources. For the public

Table 9.9 Items with mean reaching the significance level

Item	Undergraduates		Lower than bachelor's degree		With a bachelor's degree or higher		Overall			Results of selection	
	Mean	Skewness	Mean	Skewness	Mean	Skewness	Mean	Skewness	P-value	First round of selection	Second round of selection
EXP01 Factual report	4.55	-1.85	4.46	-1.45	4.66	-2.00	4.49	-1.83	0.000	③	IV +
EXP26 No false advertising	4.37	-1.12	4.44	-1.31	4.56	-1.99	4.39	-1.49	0.001	③–	I V +
EXP02 Full coverage	4.36	-1.44	4.34	-1.33	4.52	-1.29	4.36	-1.45	0.002	③–	IV +
EXP23 Not high-flown	4.36	-1.35	4.31	-1.23	4.49	-1.41	4.32	-1.43	0.002	③	IV +
EXP18 Respect for privacy	4.42	-1.34	4.28	-1.04	4.37	-1.44	4.30	-1.28	0.037	③	IV +
EXP07 Prompt and timely	4.33	-1.36	4.25	-0.97	4.39	-1.35	4.26	-1.24	0.004	③	IV +
EXP14 Government supervision	4.23	-0.93	4.12	-1.08	4.37	-1.41	4.20	-1.17	0.000	③–	IV +
EXP05 Objective without bias	4.15	-0.73	4.07	-0.76	4.30	-1.29	4.13	-0.94	0.006	③	IV
EXP12 Focus on reality	4.11	-0.81	4.07	-0.58	4.20	-0.61	4.07	-0.72	0.015	③	IV-
EXP19 Exposing wrongdoing	4.10	-0.75	4.01	-0.76	4.18	-0.98	4.04	-0.91	0.005	③	IV
EXP28 Elegant style	4.09	-0.75	3.95	-0.80	4.15	-1.30	4.03	-1.01	0.008	③	IV
EXP25 Proportion of advertising	4.00	-0.59	3.63	-0.42	3.97	-0.72	3.83	-0.64	0.000	② +	III
EXP08 In-depth	3.91	-0.66	3.80	-0.28	3.96	-0.89	3.82	-0.61	0.009	②	III
EXP10 Novel and vivid	3.43	-0.27	3.37	-0.22	3.62	-0.38	3.50	-0.33	0.012	②	III-
EXP15 Negative news	3.00	0.15	3.42	-0.40	2.97	-0.07	3.14	-0.09	0.000	① +	II
EXP16 Consistency	2.41	0.60	2.97	0.08	2.43	0.56	2.59	0.42	0.000	①	I
EXP17 High authority	2.29	0.75	2.71	0.24	2.28	0.66	2.45	0.56	0.000	①	I

Table 9.10 Comparison of the mean difference of the three groups

Dependent variab1e	(I) Group with different education backgrounds	(J) Group with different education backgrounds	Mean difference (I-J)	Standard deviation	P-value
EXP01 Factual report	Undergraduates	Lower than a bachelor's degree	0.23*	0.081	0.015
		With a bachelor's degree or higher	−0.13	0.078	0.227
	Lower than a bachelor's degree	Undergraduates	−0.23*	0.081	0.015
		With a bachelor's degree or higher	−0.37*	0.079	0.000
	With a bachelor's degree or higher	Undergraduates	0.13	0.78	0.227
		Lower than a bachelor's degree	−0.37*	0.079	0.000
EXP26 No false advertising	Undergraduates	Lower than a bachelor's degree	0.06	0.085	0.775
		With a bachelor's degree or higher	−0.24*	0.082	0.014
	Lower than a bachelor's degree	Undergraduates	−0.06	0.085	0.775
		With a bachelor's degree or higher	−0.30*	0.084	0.002
	With a bachelor's degree or higher	Undergraduates	0.24*	0.082	0.014
		Lower than a bachelor's degree	0.30*	0.084	0.002
EXP02 Full coverage	Undergraduates	Lower than a bachelor's degree	0.09	0.082	0.528
		With a bachelor's degree or higher	−0.18	0.079	0.067
	Lower than a bachelor's degree	Undergraduates	−0.09	−0.082	0.528
		With a bachelor's degree or higher	−0.28*	0.081	0.003
	With a bachelor's degree or higher	Undergraduates	0.18	0.079	0.067
		Lower than a bachelor's degree	0.28*	0.081	0.003
EXP23 Not high-flown	Undergraduates	Lower than a bachelor's degree	0.16	0.089	0.199
		With a bachelor's degree or higher	−0.15	0.086	0.210
	Lower than a bachelor's degree	Undergraduates	−0.16	0.089	0.199
		With a bachelor's degree or higher	−0.31*	0.087	0.002
	With a bachelor's degree or higher	Undergraduates	0.15	0.086	0.210
		Lower than a bachelor's degree	0.31*	0.087	0.002

(continued)

Table 9.10 (continued)

Dependent variab le	(I) Group with different education backgrounds	(J) Group with different education backgrounds	Mean difference (I-J)	Standard deviation	P-value
EXP18 Respect for privacy	Undergraduates	Lower than a bachelor's degree With a bachelor's degree or higher	0.20 −0.01	0.090 0.087	0.091 0.998
	Lower than a bachelor's degree	Undergraduates With a bachelor's degree or higher	−0.20 −0.20	0.090 0.089	0.091 0.073
	With a bachelor's degree or higher	Undergraduates Lower than a bachelor's degree	0.01 0.20	0.087 0.089	0.998 0.073
EXP07 Prompt and timely	Undergraduates	Lower than a bachelor's degree With a bachelor's degree or higher	0.20 −0.08	0.086 0.082	0.075 0.598
	Lower than a bachelor's degree	Undergraduates With a bachelor's degree or higher	−0.20 −0.28*	0.086 0.084	0.075 0.004
	With a bachelor's degree or higher	Undergraduates Lower than a bachelor's degree	0.08 0.28*	0.082 0.084	0.598 0.004
EXP14 Government supervision	Undergraduates	Lower than a bachelor's degree Lower than a bachelor's degree	0.18 −0.18	0.092 0.088	0.146 0.128
	Lower than a bachelor's degree	Undergraduates With a bachelor's degree or higher	−0.18 −0.36*	0.092 0.090	0.146 0.000
	With a bachelor's degree or higher	Undergraduates Lower than a bachelor's degree	0.18 0.36*	0.088 0.090	0.128 0.000
EXP05 Objective without bias	Undergraduates	Lower than a bachelor's degree With a bachelor's degree or higher	0.12 −0.18	0.093 0.089	0.438 0.144
	Lower than a bachelor's degree	Undergraduates With a bachelor's degree or higher	−0.12 −0.30*	0.093 0.091	0.438 0.005
	With a bachelor's degree or higher	Undergraduates Lower than a bachelor's degree	0.18 0.30*	0.089 0.091	0.144 0.005

(continued)

Table 9.10 (continued)

Dependent variab1e	(I) Group with different education backgrounds	(J) Group with different education backgrounds	Mean difference (I-J)	Standard deviation	P-value
EXP12 Focus on reality	Undergraduates	Lower than a bachelor's degree With a bachelor's degree or higher	0.14 −0.11	0.089 0.085	0.275 0.413
	Lower than a bachelor's degree	Undergraduates With a bachelor's degree or higher	−0.14 −0.26*	0.089 0.087	0.275 0.014
	With a bachelor's degree or higher	Undergraduates Lower than a bachelor's degree	0.11 0.26*	0.085 0.087	0.413 0.014
EXP19 Exposing wrongdoing	Undergraduates	Lower than a bachelor's degree With a bachelor's degree or higher	0.19 −0.11	0.093 0.090	0.115 0.480
	Lower than a bachelor's degree	Undergraduates With a bachelor's degree or higher	−0.19 −0.30*	0.093 0.092	0.115 0.004
	With a bachelor's degree or higher	Undergraduates Lower than a bachelor's degree	0.11 0.30*	0.090 0.092	0.480 0.004
EXP28 Elegant style	Undergraduates	Lower than a bachelor's degree With a bachelor's degree or higher	0.19 −0.11	0.097 0.093	0.160 0.499
	Lower than a bachelor's degree	Undergraduates With a bachelor's degree or higher	−0.19 −0.30*	0.097 0.095	0.160 0.008
	With a bachelor's degree or higher	Undergraduates Lower than a bachelor's degree	0.11 0.30*	0.093 0.095	0.499 0.008
EXP25 Controlling proportion of advertising	Undergraduates	Lower than a bachelor's degree With a bachelor's degree or higher	0.37* 0.01	0.100 0.096	0.001 0.996
	Lower than a bachelor's degree	Undergraduates With a bachelor's degree or higher	−0.37* −0.36*	0.100 0.098	0.001 0.001
	With a bachelor's degree or higher	Undergraduates Lower than a bachelor's degree	−0.01 0.36*	0.096 0.098	0.996 0.001

(continued)

Table 9.10 (continued)

Dependent variab le	(I) Group with different education backgrounds	(J) Group with different education backgrounds	Mean difference (I-J)	Standard deviation	P-value
EXP08 In-depth	Undergraduates	Lower than a bachelor's degree With a bachelor's degree or higher	0.23 −0.06	0.097 0.094	0.067 0.830
	Lower than a bachelor's degree	Undergraduates With a bachelor's degree or higher	−0.23 −0.28*	0.097 0.096	0.067 0.013
	With a bachelor's degree or higher	Undergraduates Lower than a bachelor's degree	0.06 0.28*	0.094 0.096	0.830 0.013
EXP10 Novel and vivid	Undergraduates	Lower than a bachelor's degree With a bachelor's degree or higher	0.10 −0.21	0.106 0.101	0.655 0.125
	Lower than a bachelor's degree	Undergraduates With a bachelor's degree or higher	−0.10 −0.30*	0.106 0.104	0.655 0.014
	With a bachelor's degree or higher	Undergraduates Lower than a bachelor's degree	0.21 0.30*	0.101 0.104	0.125 0.014
EXP15 Controlling negative news	Undergraduates	Lower than a bachelor's degree With a bachelor's degree or higher	−0.46* −0.05	0.122 0.118	0.001 0.897
	Lower than a bachelor's degree	Undergraduates With a bachelor's degree or higher	0.46* 0.40*	0.122 0.120	0.001 0.004
	With a bachelor's degree or higher	Undergraduates Lower than a bachelor's degree	0.05 −0.40*	0.118 0.120	0.897 0.004
EXP16 Consistency	Undergraduates	Lower than a bachelor's degree With a bachelor's degree or higher	−0.59* −0.05	0.125 0.120	0.000 0.907
	Lower than a bachelor's degree	Undergraduates With a bachelor's degree or higher	0.59* 0.53*	0.125 0.123	0.000 0.000
	With a bachelor's degree or higher	Undergraduates Lower than a bachelor's degree	0.05 −0.53*	0.120 0.123	0.907 0.000

(continued)

Table 9.10 (continued)

Dependent variable	(I) Group with different education backgrounds	(J) Group with different education backgrounds	Mean difference (I-J)	Standard deviation	P-value
EXP17 High authority	Undergraduates	Lower than a bachelor's degree	−0.54*	0.127	0.000
		With a bachelor's degree or higher	0.03	0.122	0.976
	Lower than a bachelor's degree	Undergraduates	0.54*	0.127	0.000
		With a bachelor's degree or higher	0.56*	0.125	0.000
	With a bachelor's degree or higher	Undergraduates	−0.03	0.122	0.976
		Lower than a bachelor's degree	−0.56*	0.125	0.000

Note * It means 95% chances that difference level reaches significant

the media's level and the consistency with the Party and the government are the most important, even the only factor that determines media credibility and influence. Meanwhile, propagandas are dominated by positive news, and it is believed that negative news will affect social stability and unity.

At the time of reform and opening up, the media system has changed with the social environment. The concept of the media has begun to change. More and more people realize that the media should not just be the mouthpiece of the government, but should be responsible for supervising government and caring about the public. The consistency of the media and the government will weaken people's trust in the media. Now, the media should be a "watchdog (or "gatekeeper"). Once a suspicious phenomenon is found to affect the home safety, they should "bark" to raise the host's attention. The media, like a "watchdog" of the government, should expose wrongdoing. By "demonstrating" public opinion they draw attention of relevant departments and push them to take the corresponding measures. In this concept the media should focus on negative reports.

The difference between the two concepts is: What kind of relationship should the media maintain with the government? Should the media be completely independent of the government?

The group "lower than a bachelor's degree" tend to agree on the first view, and most of them judge whether the media is credible by the authority and consistency of the media. The groups with higher education level are inclined to the second concept.

Taking "EXP16 the media is consistent with the Party and the government under all circumstances" as an example, we analyse and compare the differences among three groups in bar charts (see Figs. 9.21, 9.22, 9.23 and 9.24).

The skewness of the scores of the group of "undergraduate" and the group "with a bachelor's degree or higher" are 0.60 and 0.56 respectively, which is higher than that of the one "lower than a bachelor's degree" (0.08). It indicates that the people with higher education level are more concentrated on idea of "very unimportant"

Fig. 9.21 . *Notes* 1.0 = very unimportant; 2.0 = not important; 3.0 = slightly important; 4.0 = fairly important; 5.0 = very important

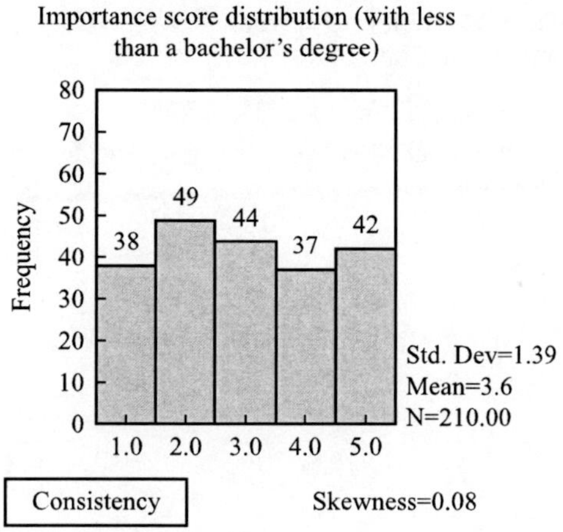

Fig. 9.22 . *Notes* 1.0 = very unimportant; 2.0 = not important; 3.0 = slightly important; 4.0 = fairly important; 5.0 = very important

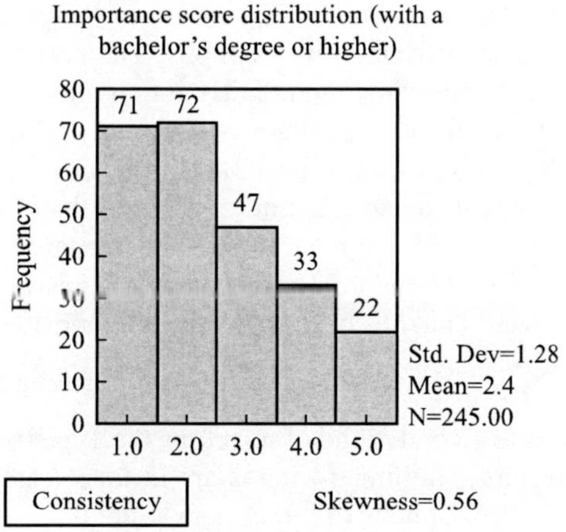

Fig. 9.23 . *Notes* 1.0 = very unimportant; 2.0 = not important; 3.0 = slightly important; 4.0 = fairly important; 5.0 = very important

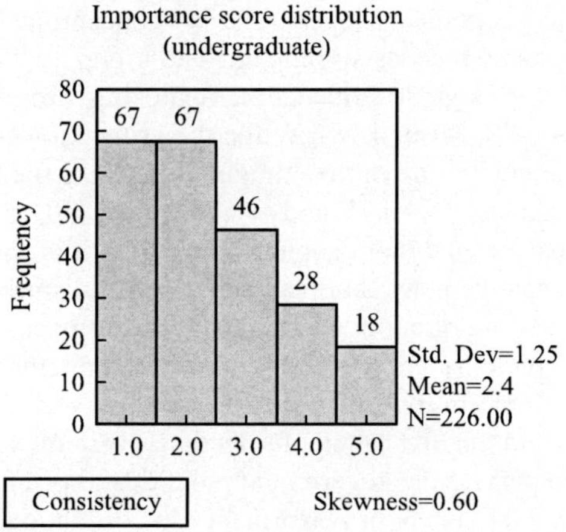

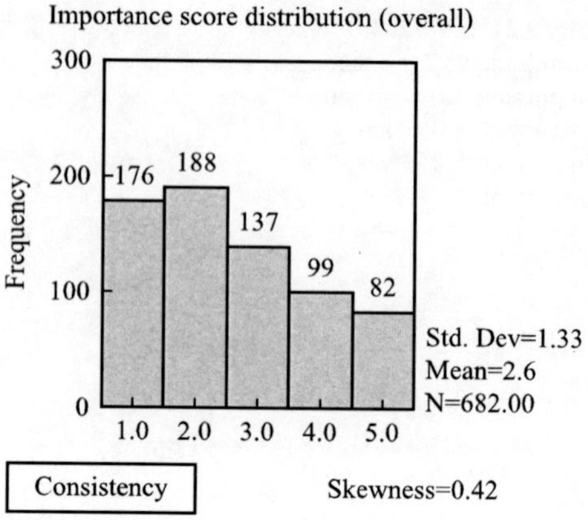

Fig. 9.24 . *Notes* 1.0 = very unimportant; 2.0 = not important; 3.0 = slightly important; 4.0 = fairly important; 5.0 = very important

and "not important". The results of the group "lower than a bachelor's degree" are evenly distributed in 5 options. The proportion of people who believe this item is "very important" and "fairly important" has reached 37.6%. For the item "EXP17 The media of high authority represent the higher-level Party and government offices", the differences are similarly significant.

But in general, the mean score of the group "lower than a bachelor's degree" on the items of "high authority" and "consistency" does not exceed "3 slightly important" (2.71 and 2.93 respectively). It means the distance between the media and the government has been recognized by most people.

② Similar opinions but different in strength.

In the second round of selection, the Type III and IV items show no obvious opposing results. The first 14 items are in these categories. The results on the 14 items are similar but have differences in strength.

For "EXP19 With the courage to expose wrongdoing and high proportion of critical reports", the general view concentrates on the 4 and 5. The difference among groups reaches significant. The group "with a bachelor's degree or higher" clearly express great influence of "exposing wrongdoing" on media credibility (skewness −0.98, Mean 4.18). While the group "lower than a bachelor's degree" and "undergraduate" are more ambiguous between the "fairly important" and "very important" (skewness −0.75 and −0.76, mean 4.01 and 4.10). In the previous analysis, we have found that people tend to think it is "not important" that the item "controlling negative news that may cause social disturbance" (skewness −0.01, mean 3.14). It means critical news has a significant positive impact on credibility, and the media which prefers reporting positive news rather than negative news will suffer the loss of credibility.

In the first round of selection, the item "EXP14 able to independently review and supervise the government policies and actions" is classified into Type ③−, for it has a relatively poor performance in correlation. Here we found the lack of consistency may be related to people's education backgrounds. Group "with a bachelor's degree or higher" tend to regard it as "very important", and the skewness reached −1.41, with an average value of 4.37. The opinions of the other two groups are skewed toward "important", but the it is less strong. This result is consistent with the previous item, "the media is consistent with the Party and the government under all circumstances". The group "lower than a bachelor's degree" who tends to agree on the importance of "consistency" do not recognise the importance of the media's "independence". The group with higher education level hold opposing views on the two items.

The scores of the three groups and the overall score on this item are shown in Figs. 9.25, 9.26, 9.27 and 9.28.

The distribution of opinions is similar for the item "EXP28 the reporting style is elegant, without vulgar news". The group "with a bachelor's degree or higher" more strongly agree its importance (skewness −1.30, mean 4.15) than the other two groups (undergraduate: skewness -0.80, mean 3.95; the group "lower than a

Fig. 9.25 . *Notes* 1.0 = very unimportant; 2.0 = not important; 3.0 = slightly important; 4.0 = fairly important; 5.0 = very important

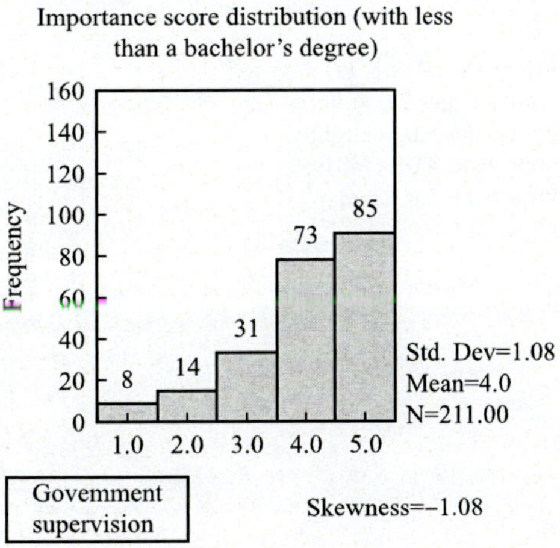

Fig. 9.26 . *Notes* 1.0 = very unimportant; 2.0 = not important; 3.0 = slightly important; 4.0 = fairly important; 5.0 = very important

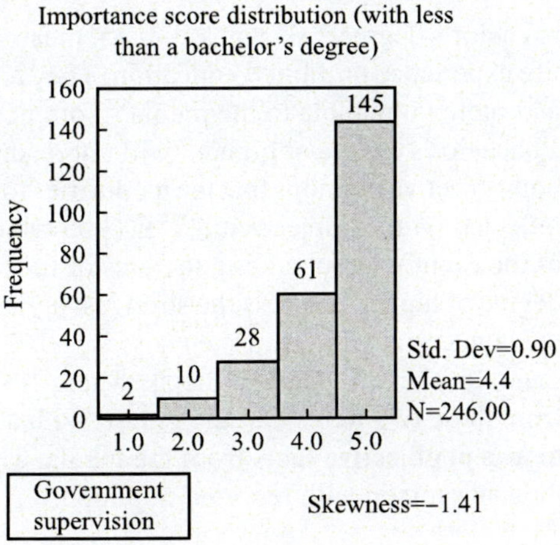

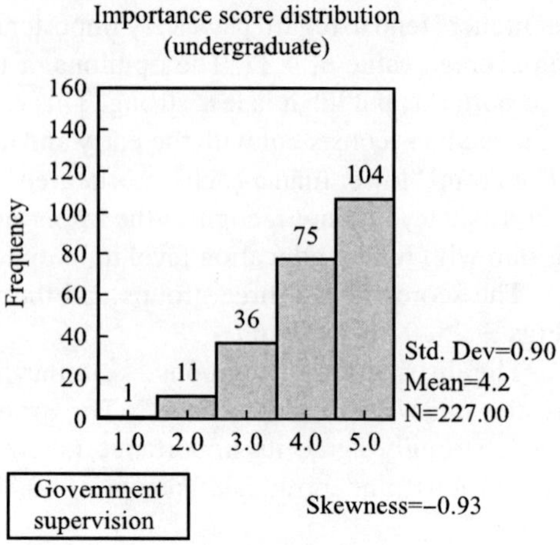

Fig. 9.27 . *Notes* 1.0 = very unimportant; 2.0 = not important; 3.0 = slightly important; 4.0 = fairly important; 5.0 = very important

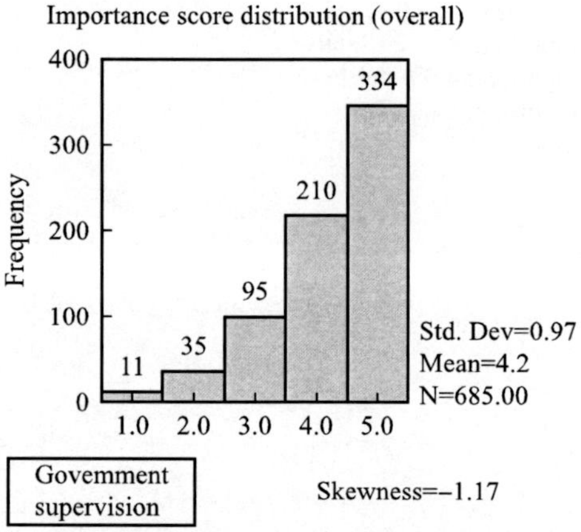

Fig. 9.28 . *Notes* 1.0 = very unimportant; 2.0 = not important; 3.0 = slightly important; 4.0 = fairly important; 5.0 = very important

bachelor's degree": skewness −0.75, mean 4.09). This is probably due to the lack of life experience or limited education. They are not good at identifying publicity tricks and more vulnerable to the media's comment. Relatively speaking, the group "with a bachelor's degree or higher" will not readily accept or even feel disgusted with the non-objective opinions that the media tries to impose on readers. The difference is also reflected on the "objectivity" ("EXP05 Objectively present the original appearance of the event without adding the bias of the reporter"). The group "with a bachelor's degree or higher" express the strongest agreement (skewness −1.29, mean 4.30). For the "undergraduate", the skewness is −0.73 and mean is 4.15; for the group "lower than a bachelor's degree" the skewness is −0.76 and mean is 4.07. It can be seen from these two items that the group "with a bachelor's degree or higher" is inclined to accept objective facts from the media. The other two groups are less resistant to the news mixed with reporter's opinion.

The group "with a bachelor's degree or higher" gave the highest score at the four items related to journalistic writing skills, "EXP07 news reports are prompt and timely", "EXP08 able to provide in-depth analysis and interpretation of complex news events", "EXP10 the news layout or program is novel, vivid and unique" and "EXP23 The news reports tell the truth, not using high-flown rhetoric". When evaluating media credibility, they more value the skills and forms of news than the group with lower education level.

A certain pattern exists in the opinion of three groups on the first 14 items. The group "with a bachelor's degree or higher" have the strongest view towards most of items. The "undergraduate" is second to it while the opinion of the group "lower than a bachelor's degree" is relatively weaker. Only three items are exceptions to this pattern.

For "EXP26 no false advertising", the group "with a bachelor's degree or higher" have the strongest view. The second is the group "lower than a bachelor's degree" instead of "undergraduate". It may be attributed that undergraduates are still in college and have less chance to be troubled by false advertising. Correspondingly, "undergraduates" expressed the strongest agreement to the "EXP25 The proportion of advertising is appropriate", while the other two groups with more life experience are more neutral to it. The group of undergraduates give highest score for importance to "EXP18 respect for people's privacy", followed by the group "with a bachelor's degree or higher" and "lower than a bachelor's degree".

9.4.3.4 Correction of Classification Based on Differences Between Groups

It can be found that various opinions among groups with different education backgrounds may be related to the concepts of the media, knowledge, information interpretation, life experience, etc. Among them, the difference caused by the concept of the media is worthy of attention. The remaining differences have no significant impact on the classification. Considering that highly qualified proportion of the whole country is lower than that of the sample, we use "+" and "-" to mark the group "lower than a bachelor's degree". If the mean score is lower than the average, the results is marked with "-" to reduce the level of the item in selection results. If the mean score is higher than the overall score, it is marked with "+" to increase level of the item in selection results. Items with no differences between groups are maintained the original results (see Table 9.11).

Table 9.11 Correction of the importance of items based on differences between groups

Item	Overall sample		Results of selections		
	Mean	Skewness	First round of selection	Second round of selection	Correction
EXP01 Factual report	4.49	-1.83	③	IV +	--
EXP20 Public view	4.43	-1.58	③	IV +	
EXP26 No false advertising	4.39	-1.49	③ –	IV +	+
EXP22 Equal manner	4.37	-1.41	③	IV +	
EXP02 Full coverage	4.36	-1.45	③–	IV +	--
EXP23 Not high-flown	4.32	-1.43	③	IV +	--
EXP11 Correcting errors	4.31	-0.92	③	IV	
EXP18 Respect for privacy	4.30	-1.28	③	IV +	--
EXP21 Concerned about the disadvantaged groups	4.28	-1.26	③	IV +	
EXP04 Accurate report	4.26	-1.14	③	IV +	
EXP07 Prompt and timely	4.26	-1.24	③	IV +	--
EXP06 Balancing without bias	4.26	-1.23	③	IV +	
EXP14 Government supervision	4.20	-1.17	③–	IV +	--
EXP05 Objective without bias	4.13	-0.94	③	IV	--
EXP12 Focus on reality	4.07	-0.72	③	IV-	--
EXP13 Sharp perspective	4.06	-0.72	③	IV-	
EXP19 Exposing wrongdoing	4.04	-0.91	③	IV	--
EXP28 Elegant style	4.03	-1.01	③	IV	--
EXP29 Independent editing	3.96	-0.87	② +	IV-	
EXP03 Complete coverage	3.91	-0.74	② +	IV-	
EXP25 Controlling proportion of advertising	3.83	-0.64	② +	III	--
EXP08 In-depth	3.82	-0.61	②	III	--
EXP32 Positive review	3.70	-0.62	②	III	
EXP27 No paid news	3.61	-0.35	②	III-	
EXP10 Novel and vivid	3.50	-0.33	②	III-	--
EXP24 Reducing conference news	3.43	-0.28	①	III-	
EXP31 Large scale	3.38	-0.34	②–	III-	
EXP09 Exclusive	3.19	0.01	②–	II +	
EXP15 Controlling negative news	3.14	-0.09	①	II	
EXP30 Long history	2.82	0.19	①	II	+
EXP16 Consistency	2.59	0.42	①	I	+
EXP17 High authority	2.45	0.56	①	I	+

9.4.4 *Summary of Selection Results of Media Credibility Evaluation Criteria*

9.4.4.1 The Overall Importance of Each Item

After the two rounds of selection and the corrections, we have a basic understanding of what factors Chinese people value in evaluating media credibility. Selections and correction are based on different standards. The classification from the results of the selection is set to be represented by the corresponding value, that is, ① = 1, ② = 2, I = 1, II = 2 and so on, while "+" is equivalent to "+0.5", "−" is equivalent to a "−0.5". The results of selection and correction are added together to obtain a total score of each item. The minimum of the total score is 1.5 and the maximum is 8. The total value can be used as a reference value for media credibility evaluation. The selection results of the 32 items of "Importance Scale" are shown in Table 9.12 (the table are arranged in descending order according to the total result).

The added-up score of importance shows the first 17 items, especially the first 11 items, can be considered as important criteria for credibility evaluation; the last 5 items are not suitable. The other 10 in the middle belong to the non-important standards.

9.4.4.2 Factor Analysis of Media Credibility Evaluation Criteria

Further analysis of the public's view on credibility can be conducted based on the above results.

The 32 items in the Importance Scale are designed based on six categories;

Journalistic professionalism (the basic professional requirements of journalism)
- Social concern (mainly refers to the interest of the public)
- Media Ethics (the media is not driven by benefit to harm the public interest)
- Journalistic skills (skills in the form of presenting events)
- The relationship between the media and the government
- External image

Items are rearranged according to its category with the results of selections. The importance of each item can be seen from the results (see Table 9.13).

In terms of "relationship between the media and the government", the item "EXP14 able to independently review and supervise the government policies and actions" shows the contrary to the other three items. It means that the consistency with the government is in contrast to government supervision and media credibility. Scores at "media's external image" is generally low, which is deemed as little relation to media credibility.

Overall, the public think that "social concern" and "journalistic professionalism" are the most important factors for media credibility and "media ethics" and "journalistic skills" are secondary.

Table 9.12 The overall importance score of each item

Ranking	Item		Mean	First round of selection	Second round of selection	Correction	Added-up (range 1.5 -7.5)
1	EXP20	Public view	4.43	③	IV +		7.5
2	EXP26	No false advertising	4.39	③ –	IV +	+	7.5
3	EXP22	Equal manner	4.37	③	IV +		7.5
4	EXP21	Concerned about the disadvantaged groups	4.28	③	IV +		7.5
5	EXP04	Accurate report	4.26	③	IV +		7.5
6	EXP06	Balancing without bias	4.26	③	IV +		7.5
7	EXP01	Factual report	4.49	③	IV +	--	7
8	EXP23	Not high-flown	4.32	③	IV +	--	7
9	EXP11	Correcting errors	4.31	③	IV		7
10	EXP18	Respect for privacy	4.30	③	IV +	--	7
11	EXP07	Prompt and timely	4.26	③	IV +	--	7
12	EXP02	Full coverage	4.36	③–	IV +	--	6.5
13	EXP14	Government supervision	4.20	③ –	IV +	--	6.5
14	EXP05	Objective without bias	4.13	③	IV	--	6.5
15	EXP13	Sharp perspective	4.06	③	IV-		6.5
16	EXP19	Exposing wrongdoing	4.04	③	IV	--	6.5
17	EXP28	Elegant style	4.03	③	IV	--	6.5
18	EXP12	Focus on reality	4.07	③	IV-	--	6
19	EXP29	Independent editing	3.96	② +	IV-		6
20	EXP03	Complete coverage	3.91	② +	IV-		6
21	EXP25	Controlling proportion of advertising	3.83	② +	III	--	5
22	EXP32	Positive review	3.70	②	III		5
23	EXP08	In-depth	3.82	②	III	--	4.5
24	EXP27	No paid news	3.61	②	III-		4.5
25	EXP10	Novel and vivid	3.50	②	III-	--	4
26	EXP31	Large scale	3.38	②–	III-		4
27	EXP09	Exclusive	3.19	②–	II +		4
28	EXP24	Reducing conference news	3.43	①	III-		3.5
29	EXP30	Long history	2.82	①	II	+	3.5
30	EXP15	Controlling negative news	3.14	①	II		3
31	EXP16	Consistency	2.59	①	I	+	2.5
32	EXP17	High authority	2.45	①	I	+	2.5

Table 9.13 Categories before study and the results of selections

Category	Item	Mean	Added-up score
Social concern	EXP20 Public view	4.43	7.5
	EXP22 Equal manner	4.37	7.5
	EXP21 Concerned about the disadvantaged groups	4.28	7.5
	EXP23 Not high-flown	4.32	7
	EXP18 Respect for privacy	4.3	7
	EXP13 Sharp perspective	4.06	6.5
	EXP19 Exposing wrongdoing	4.04	6.5
	EXP12 Focus on reality	4.07	6
	EXP24 Reducing conference news	3.43	3.5
Journalistic professionalism	EXP04 Accurate report	4.26	7.5
	EXP06 Balancing without bias	4.26	7.5
	EXP01 Factual report	4.49	7
	EXP02 Full coverage	4.36	6.5
	EXP05 Objective without bias	4.13	6.5
	EXP03 Complete coverage	3.91	6
Media Ethics	EXP26 No false advertising	4.39	7.5
	EXP28 Elegant style	4.03	6.5
	EXP29 Independent editing	3.96	6
	EXP25 Controlling proportion of advertising	3.83	5
	EXP27 No paid news	3.61	4.5
Journalistic skills	EXP11 Correcting errors	4.31	7
	EXP07 Prompt and timely	4.26	7
	EXP08 In-depth	3.82	4.5
	EXP10 Novel and vivid	3.5	4
	EXP09 Exclusive	3.19	4
External image	EXP32 Positive review	3.7	5
	EXP31 Large scale	3.38	4
	EXP30 Long history	2.82	3.5
Relationship between the media and the government	EXP14 Government supervision	4.2	6.5
	EXP15 Controlling negative news	3.14	3
	EXP16 Consistency	2.59	2.5
	EXP17 High authority	2.45	2.5

Then, we decide to maintain the "government supervision" item, and delete the six items with lower score in "external image" and "relationship between the media and the government", and delete "reducing conference news" (not suitable for evaluation). We perform exploratory factor analysis on the "Importance Scale". By using principal components, maximum variance rotation (VARIMAX), and extracting common factors with values greater than 1, five factors are obtained (see Table 9.14).

The fifth factor corresponds to "public view", and "concerned about the disadvantaged groups" has a close relationship with the second factor (the factor loading is exceeded 0.4). It is difficult to explain that the three items (exposing wrongdoing,

Table 9.14 The first factor analysis of Importance scale

Item		Factor 1	Factor 2	Factor 3	Factor 4	Factor 5
EXP01	Factual report	0.721			0.201	0.234
EXP02	Full coverage	0.707			0.191	0.258
EXP04	Accurate report	0.641	0.321		0.139	
EXP05	Objective without bias	0.597	0.296	0.108	0.152	
EXP06	Balancing without bias	0.596	0.403	0.116		
EXP03	Complete coverage	0.522	0.189	0.296	0.117	0.111
EXP14	Government supervision	0.517	0.113			0.39
EXP22	Equal manner	0.227	0.673		0.273	0.218
EXP23	Not high-flown	0.241	0.637		0.258	0.187
EXP07	Prompt and timely	0.292	0.495	0.379		
EXP12	Focus on reality	0.142	0.494	0.473	0.116	
EXP11	Correcting errors	0.257	0.486	0.289	0.203	0.107
EXP18	Respect for privacy	0.272	0.44		0.204	0.242
EXP09	Exclusive			0.775		
EXP10	Novel and vivid			0.769	0.139	0.117
EXP08	In-depth	0.14	0.45	0.548		
EXP13	Sharp perspective	0.122	0.492	0.542	0.138	0.137
EXP27	No paid news	0.155		0.28	0.673	
EXP28	Elegant style	0.201	0.201		0.664	
EXP25	Controlling proportion of advertising		0.139		0.659	0.133
EXP26	No false advertising	0.156	0.391		0.621	
EXP29	Independent editing	0.288	0.11	0.147	0.533	0.203
EXP19	Exposing wrongdoing	0.224		0.213	0.133	0.734
EXP20	Public view	0.187	0.419	0.142	0.129	0.606
EXP21	Concerned about the disadvantaged groups		0.505		0.114	0.588

Note The cumulative variance ratio of the factor is 53.9%

public view and concerned about the disadvantaged groups) are independent as one factor. With limited extraction of four factors, the three items of the fifth factor are attributed to the second factor. Although the contribution of the factor decreases slightly, the results obtained are more reasonable. According to the content of the items included in each factor, we name the first four factors as "journalistic professionalism", "social concern", "journalistic skills" and "media ethics" respectively. With reference to the previous analysis of the importance of each item, we find the importance of the four factors for the evaluation of media credibility (see Table 9.15).

Compared with the classification we used previously, it can be found that the results of the factor analysis are basically consistent with the original assumptions, and only 4 items don't match.

The item "EXP14 able to independently review and supervise the government policies and actions" is classified into "journalistic professionalism". It shows that the public regards the media's review and supervision of the government as basic values of journalism, like "factual", "objective" and "accurate".

The item, "EXP13 good at finding problems and having a sharp perspective" is classified into the factor of "journalistic skills" instead of "social concern". The factor loading of this item on "social concern" is high, indicating that it is related to both aspects. The initial assumption of this item is that "the media have a sharp perspective" so that they are "good at finding problems" in the society. The statistical analysis reveals that "sharp perspective" is necessarily related to the issue related to the public, but the "sharp perspective" can be viewed as a journalistic skill.

It is used as "journalistic skills" before study that "EXP11 Apologise and correct wrong reports once they have been found", but the factor analysis results classified it into "social concern". Thus it shows that the public view the correcting errors as a sign of taking responsibility. Its factor loading is close to 0.4 in "journalistic skills", so it is relevant to the "journalistic skills".

The item "EXP12 the news covers the most realistic and urgent issues in social development" is reclassified into "journalistic skills" instead of "social concern" before study, but its factor loading exceeds 0.4 on the "social concern", so its connotations are related to the two factors.

9.4.5 Analysis of the Characteristics of China's Media Credibility Evaluation Criteria

9.4.5.1 The Media that Speak for the People Are More Credible

After selections and factor analysis, we conclude the criteria for media credibility evaluation, which contain 4 dimensions—"journalistic professionalism", "social concern", "journalistic skills" and "media ethics". The mean value of the added up scores of the four dimensions are 6.79, 7.14, 5.33, and 5.9, which represent the

Table 9.15 The second factor analysis and importance of factor

Dimension	Item		Factor 1	Factor 2	Factor 3	Factor 4	Added-up scores	Factor importance
Journalistic professionalism	EXP01	Factual report	0.744	0.163		0.186	7	6.79
	EXP02	Full coverage	0.734			0.162	6.5	
	EXP04	Accurate report	0.632	0.217	0.113	0.178	7.5	
	EXP06	Balance without bias	0.592	0.315	0.157	0.139	7.5	
	EXP05	Objective without bias	0.58	0.159	0.141	0.196	6.5	
	EXP14	Government supervision	0.559	0.272			6.5	
	EXP03	Complete coverage	0.53	0.165	0.313	0.123	6	
social concern	EXP21	Concerned about the disadvantaged group	0.117	0.729	0.121		7.5	7.14
	EXP22	Equal manner	0.246	0.66		0.313	7.5	
	EXP20	Public view	0.254	0.655	0.175		7.5	
	EXP23	No bureaucratic tone	0.256	0.616		0.299	7	
	EXP18	Respect for privacy	0.296	0.478		0.219	7	
	EXP11	Correcting errors	0.263	0.423	0.336	0.236	7	
	EXP19	Exposing wrongdoing	0.31	0.413	0.206		6.5	
Journalistic skills	EXP09	Exclusive			0.767		4	5.33
	EXP10	Novel and vivid			0.764	0.115	4	
	EXP08	In-depth	0.14	0.356	0.59		4.5	
	EXP13	Sharp perspective	0.132	0.435	0.587	0.165	6.5	
	EXP12	Focus on reality	0.145	0.411	0.52	0.151	6	
	EXP07	Prompt and timely	0.288	0.385	0.427	0.124	7	
Media ethics	EXP27	No paid news	0.155		0.28	0.667	4.5	5.9
	EXP28	Elegant style	0.213	0.191	0.1	0.665	6.5	
	EXP26	No false advertising	0.154	0.311		0.657	7.5	
	EXP25	Controlling proportion of advertising		0.172		0.647	5	
	EXP29	Independent editing	0.313	0.165	0.154	0.512	6	

Note The factor cumulative variance ratio is 49.7%

importance of each dimension in the credibility evaluation (see Table 9.15). It indicates the public values the "social concern" most in evaluating media credibility, i.e. the social responsibility of the media, humanistic care, and closeness to the people. As long as it is related to "the public interest", people will pay great attention to it no matter what their education backgrounds are. The role of the media is the voice of the people, which is demand of the public at this stage. The media that can satisfy such demand will win the trust of the public. This can be seen in many cases, such as *Oriental Horizon*, a popular documentary television programmes on CCTV; and *Dahe Daily*, a newspaper which gained popularity by series reports of "Zhang Jinzhu case". The rise of such media reflects the public's trust in them.

Now more and more media has shifted their attentions to the public, but people commonly agree on the view "the media that speaks for the common people is more credible".

9.4.5.2 "Media Professionalism" Has Given Way to "Social Concern"

The results indicate that the "professionalism" of the media, (factual, accurate, complete, objective, comprehensive, etc.), has occupies a very important proportions in the evaluation of media credibility. The mean of the added-up values reached 6.86. It is consistent with the results of credibility studies in the United States. It has become a basic requirement for the media.

However, it is worth noting that the research on media credibility in other countries mainly uses "professionalism requirement" as the criteria for credibility evaluation. In many cases, "social concern" is not even included in media credibility scale. In China, the analysis shows that the importance of "social concern" is higher than that of "professionalism", and this tendency is more pronounced among people with lower education level. Among the top 17 items, the rank of "accuracy", "balance", "truth" and "full coverage" lags behind "public view", "equal manner", "concerned about the disadvantaged groups". "objective without bias", the important professionalism requirement in the Western concept ranks only 14th, and the majority lower than Bachler's degree evaluate "objective" much lower than that of the people with higher education level. The top 17 items are shown in Table 9.16 (ranking according to "added-up scores" and "overall mean" in descending order).

In reality there are reports that violate objectiveness and fairness but receive big support from the public. The most typical is the report of the judicial trial of "Zhang Jinzhu case", a police officer who attacked two people because of drunk driving, one seriously injured and other died. The sentence of Zhang was largely affected by the public opinion created by the media. In the other cases on the internet, such as "Gang leader Liu Yong case" and "mysterious death of a music teacher Huang Jing", the victims are the disadvantaged people but the accused with special political backgrounds. Any reports defending for the accused was regarded incredible, but the blame to the accused (no matter whether it is objective or has enough evidence) would be trusted.

Table 9.16 Top 17 items of importance

Ranking	Item	Overall Mean	First round of selection	Second round of selection	Correction	Added-up scores (range 1.5 -7.5)
1	EXP20 Public view	4.43	③	IV +		7.5
2	EXP26 No false advertising	4.39	③–	IV +	+	7.5
3	EXP22 Equal manner	4.37	③	IV +		7.5
4	EXP21 Concerned about the disadvantaged groups	4.28	③	IV +		7.5
5	EXP04 Accurate report	4.26	③	IV +		7.5
5	EXP06 Balancing without bias	4.26	③	IV +		7.5
7	EXP01 Factual report	4.49	③	IV +	--	7
8	EXP02 Full coverage	4.36	③–	IV +	--	6.5
9	EXP23 Not high-flown	4.32	③	IV +	--	7
10	EXP11 Correcting errors	4.31	③	IV		7
11	EXP18 Respect for privacy	4.30	③	IV +	--	7
12	EXP07 Prompt and timely	4.26	③	IV +	--	7
13	EXP14 Government supervision	4.20	③ –	IV +	--	6.5
14	EXP05 Objective without bias	4.13	③	IV	--	6.5
15	EXP13 Sharp perspective	4.06	③	IV-		6.5
16	EXP19 Exposing wrongdoing	4.04	③	IV	--	6.5
17	EXP28 Elegant style	4.03	③	IV	--	6.5

From the survey we fousnd that the public had some special understanding of the "objectiveness" of news. The survey question is as follows.

For the media using the power of public opinion to interfere in trial process, do you think it is reasonable?

1. unreasonable, the media should not go beyond its responsibility
2. *completely reasonable*
3. *at* the current stage of China, the phenomenon is reasonable.
4. *not clear.*

In the 692 samples, 33.3% of the respondents chose "unreasonable". 45.6% of respondents think it is reasonable at this stage. There are still 13.7% choosing "completely reasonable". The rest 7.4% regard it as "unclear". In other words, nearly 60% of the respondents accept the media's interference in the judicial process.

Then, the samples are divided into two groups based on education backgrounds—"lower than a bachelor's degree" and "undergraduate or with a higher degree". Cross-analysis is conducted with "whether the intervention of justice is reasonable." The results show that the opinion has a significant correlation with the education backgrounds. The group with higher educational level is more inclined to choose "unreasonable" and "reasonable to some extent. While the group with lower education level choose "completely reasonable" and "not clear". The cross analysis is shown in Table 9.17.

Results show nearly half of the respondents believe it "reasonable at current stage" even though it is not the responsibility of the media.

Then, what is "the current stage"? This is a fairly complicated issue. Since this research scope is limited to the credibility, the question is designed to investigate the major public institutions and departments.

> How much do you trust the following institutions or departments? Please rate them with 0-10 based on your impressions and feelings. A minimum of 0 points indicates complete distrust, the maximum of 10 points indicates complete trust, and 6 points means pass. The more you trust, the higher you score.

Results show that at current stage of China the credibility of public institutions is generally sluggish, mainly between 6 and 7, and the media is above average (6.64). Lower score is given by groups with more life experience, "lower than a bachelor's degree" and "with a bachelor's degree or higher", especially the latter one giving the score "not pass". The college students of 18–21 years old are more optimistic towards the credibility (see Table 9.18).

The results show that public institutions couldn't obtain high trust from people, and the public have relatively high expectations for the media. The expectation may go beyond the power of the media itself sometimes (for example, the media is expected to make justice prevail), so that the media may interfere with the justice in order to "gain public support". It has been debated for a long time whether the media should be reporters, interpreters or critics. Often, during times of social change, they are expected to be engaged in more social activities. For example, in the 1960s there was a lot of social turmoil in the United States. Many people took a skeptical and critical attitude towards the role of the media as a "neutral gatekeeper". They believed that the media should take more social responsibilities and advocated that journalists should support the disadvantaged groups. Therefore, the public expectations for the media and their credibility are a reflection of social reality in China and the background of research on media credibility.

9.4.5.3 "Media Ethics" and "Journalistic Skills" Become the Secondary Criteria for Credibility Evaluation

The importance of "media ethics", and "journalistic skills" are significantly lower than "social concern" and "professionalism", belonging to the secondary criteria.

The interest-driven behaviours that damages the public interest, such as news with excessive descriptions of violence, the proliferation of advertisements and paid news, are becoming increasingly prominent with the competition in the media market, posing a threat to media credibility.

Table 9.17 The cross analysis of "different groups" and "justice intervention of the media"

Item			Whether the intervention of justice is reasonable				Overall
			Unreasonable	Completely reasonable	Reasonable	Unclear	
Group with different education Backgrounds	Undergraduate or with a higher degree	Sample size	172	60	221	21	474
		Predictive value	157.8	65.2	216.1	35.0	474.0
		Proportion	36.3%	12.7%	46.6%	4.4%	100%
	Lower than a bachelor's degree	Sample size	58	35	94	30	217
		Predictive value	72.2	29.8	98.9	16.0	217.0
		Proportion	26.7%	16.1%	43.3%	13.8%	100%
Total		Sample size	230	95	315	51	691
		Predictive value	230.0	95.0	315.0	51.0	691.0
		Proportion	33.3%	13.7%	45.6%	7.4%	100%

Note $x^2 = 23.547$; $df = 3$; $C = 0.185$; $P = 0.000$

Table 9.18 Credibility evaluation of major social institutions and departments

Institutions and departments	Lower than a bachelor's degree	With a bachelor's degree or higher	Undergraduates	Overall
Academia	6.92	6.22	7.63	7.05
Military	7.56	6.22	7.31	7.00
People's Congress	7.49	6.12	7.23	6.92
Education	6.90	6.27	7.10	6.74
Court	7.29	5.87	7.02	6.69
The media	7.05	6.21	6.73	6.64
Notary Department	6.75	5.84	6.98	6.51
Government	6.80	5.92	6.81	6.49
Qualification Inspection Departments	6.69	5.74	6.78	6.38
Chinese People's Political Consultative Conference	6.57	5.82	6.75	6.36
Large companies	6.17	5.83	6.57	6.18
Public security agencies	6.53	5.39	6.63	6.16
Medical departments	6.27	5.67	6.40	6.10
Religious communities	5.51	5.92	6.36	5.94
Average	6.75	5.96	6.88	6.51

The survey shows that the attentions attached to the media ethics is related to people's education backgrounds. Those with higher education level express a stronger feeling towards it. It is also related to the public's ability to identify the news hype and paid news. The questionnaire designed for such ability.

> There is some news called "advertising news", "paid news", and "soft-sell advertising", which are actually promotion of products or companies. Can you identify them easily?

Respondents choosing "can easily recognize" account for 57.7%; those who "can't recognize" account for 26.4%. 15.9% of the respondents tell that they "never notice this phenomenon", which means that they may even do not know "such news" (see Table 9.19).

Table 9.19 Recognition of paid-news

Item	Sample size	Valid percentage (%)
Can easily recognize	395	57.7
Can't recognize	181	26.4
Never notice	109	15.9
Overall	685	100

Most of the respondents in this survey are ones with a bachelor's degree or higher. The data analysis reflects the ability to identify the paid news is related to the attitudes towards the "paid news". The mean to "no paid news" given by people who can identify "paid news" is 3.70. For those "cannot recognize" paid news, the mean is 3.49 (the difference of the mean value between the two groups is lower than 0.05, and it reaches a significant level).

Paid news is not easy to stop. We might need to improve people's ability to identify this kind of news. When people realize that the media irresponsibly hype the news and publish paid news for their own interests, the importance of "media ethics" in evaluating media credibility will increase accordingly.

However, the response to the "false advertising" is very consistent and strong (the mean is 4.39, ranked 2nd among 32 items). The false advertising could be fatal to media credibility.

The performance on the reporting skills has effect on the mean credibility (the mean is roughly around 3.7), and are more valued by the group with high education level. In terms of journalistic skills, the respondents emphasize the "timely" (with mean score 4.26), which is much higher than other skills. The ability to inform the public in a timely manner greatly affects media credibility.

Due to restrictions on the content of news in China, the skill of presenting events is more important. Other skills, such as in-depth, exclusiveness and novel form, have a positive impact on media credibility.

9.4.5.4 The Biggest Difference in the Relationship Between the Media and the Government

From the analysis, it can be found that items including external image (positive review, a long history, and a large scale), the relationship with the government (consistency, high authority, controlling negative news) and "reducing conference news" are not suitable as evaluation criteria, but they are used by around 10–30% people to evaluate media credibility, especially for people with relatively lower education level. Most of the respondents in this survey are those with high education level. Therefore, this percentage should be higher in the overall Chinese population.

It is not strange if there is only a difference in importance. Since people have very personal criteria, some people think that it is not so important and some people think it is very important. Research on China's media credibility should start with the most important and well-received aspects. Bu if there is a divergence of opinions, the complexity of the research will be increased. For example, for the item "*the media should be consistent with the Party and the government under any circumstances*", the proportions of "1 very unimportant" and "2 not important" are 25.4% and 27.1% respectively; while the proportions of "5 very important" and "4 fairly important" are 11.8% and 14.3% respectively. To address this issue, one more question is designed.

If the media is government-owned, will you trust it more or less?

Table 9.20 Cross-analysis results of "groups with different education backgrounds" and "change of trust" Unit: %

Item	Increase trust	Decrease trust	Depend on the specific situation	Unclear	Total
Undergraduate	2.4	50.7	46.0	0.9	100
Lower than a bachelor's degree	10.3	38.4	34.0	17.2	100
With a bachelor's degree or higher	4.2	46.0	44.7	5.1	100
Total	5.5	45.2	41.8	7.5	100

Note $N = 651$; $x^2 = 60.397$; $df = 6$; $C = 0.026$; $P = 0.000$

Respondents who think "decrease trust" and "depend on the specific situation" account for the highest proportion, 45.2% and 41.8% respectively. 5.5% think it will "increase trust". 7.5% regards it as "unclear".

Cross-analysis shows that difference among people with different education backgrounds have reach significant. The proportion "increase trust" and "unclear" in the group lower than a bachelor's degree is obviously higher than the other two groups, while undergraduates tend to "decrease trust" (see Table 9.20).

9.5 Summary

In this chapter a data-based analysis of Chinese public's expectations for the media and the corresponding evaluation criteria are conducted based on the survey of Media Expectation Scale (also Importance Scale). The criteria for evaluating media credibility are selected in this process. The results, as a vital basis, reveal the unique characteristics of China's media credibility. These unique issues would be addressed in both theory research and practical operation of China's media. Furthermore, education backgrounds lead to the differences in the concepts of news and credibility evaluation criteria, and it is particularly worthy of attention.

Chapter 10
Development of China's Media Credibility Assessment Scale

In Chap. 9 we selected criteria for evaluating credibility from the perspective of importance, and analysed the characteristics of credibility judgments of the Chinese people. The results are important to the development of the media credibility index. In addition to the importance, there are other requirements of the credibility assessment index, such as the reliability and validity, clarity and simplicity of the scale. Thus, the above items recognized as "important" or "very important" by the public cannot be used directly as indexes of the credibility assessment scale. They need to be further selected in terms of the practical effect in application. Therefore, a new scale, "Media Credibility Assessment Scale", is designed to further explore the development of China's Credibility Assessment Scale.

10.1 Design of the Assessment Scale of China's Media Credibility

The "Importance Scale" based on the previous data is mainly to examine which aspects the public would consider in evaluating media credibility. The "Media Credibility Assessment Scale" (hereinafter referred to as the Assessment Scale) is used to examine the media credibility. The objects of the two scales are different, but they are designed to support the development of the China's media Credibility Assessment scale. The "Importance Scale" explores the public's expectations and needs, while the "Assessment Scale" is to examine whether these expectations and needs have been satisfied. We have discussed the essence and concepts of media credibility specifically, and proposed that media credibility is concerned about whether the public's expectations are met. The public assessment of the credibility is the measurement of credibility. The "Assessment Scale" is consistent with the "evaluation scale".

However, the main purpose of using the "Assessment Scale" in this section is not to measure the media credibility, but to use the evaluation results to develop a scientific and suitable assessment scale. This chapter revolves around the "Assessment Scale",

G. Yu, *Research on the Communication Effects and Mass Media Credibility in China*,
https://doi.org/10.1007/978-981-19-6242-4_10

through the analysis of its reliability, validity test, exploratory and confirmatory factors, and then combined with the "Importance Scale" to develop the Assessment Scale.

The item descriptions of the "Assessment Scale" are consistent with the "Importance Scale". Considering that "Assessment Scale" evaluates the overall situation of the China's media, we delete 4 items from "Importance Scale" which only suits for a single medium. They are listed as following;

The media of high authority represent the high-level Party and government offices. (EXP17)

The media has a long history. (EXP30)

The media has a large scale circulation and volumes. (EXP31)

The media are well received by many people. (EXP32)

In addition, the results show that first two items are not suitable as media credibility evaluation criteria. The importance of the latter two items is low. Therefore, the deletion of the four will not pose much negative influence on the importance of the "Assessment Scale".

For remaining 28 items, both the content and number would be remained, and the only change is the name of variable, which is "JUD" (short for judgment) instead of "EXP". The description of the 28 items can be found in Table 9.3.

Another difference between the two scales is that "Assessment Scale" adopts a 10-point system that the Chinese people are more accustomed to, but the "Importance Scale" uses 1 to 5 points to represent different importance. The questionnaire of the "Assessment Scale" is designed as followings.

In general, how does China's media do on the items listed above? Please rate its performance with 0–10 points based on your perceptions.

(The minimum 0 points means that it has done poorly, and the maximum 10 points means that it has done very well; 6 points means pass. The better it does, the higher score you give).

10.2 The Fourth Selection of Indexes of China's Media Credibility Assessment Scale

The first round of selection was based on the importance rating score in the "Importance Scale". The second round of selection was from distribution of scores, and results was corrected according to the differences among groups with different education backgrounds. The following analysis and selection of indexes of credibility evaluation scale on the basis of the "Assessment Scale" can be called the fourth round of selection.

10.2.1 Methods of the Fourth Round of Selection

10.2.1.1 The Basic Principles of Developing a Scale

There are two fundamental principles for developing a scale– "accurate" and "concise". The analyses conducted for the "Importance Scale" was mainly from the accuracy, examining whether each item is suitable as a criterion and its importance. This section is a comprehensive analysis of the scale development from the perspectives of "accuracy" and "conciseness." The conciseness is not just simple. If the scale it is too simple, it will affect the accuracy of the scale. Therefore, it is necessary to analyse the actual situation according to these two principles.

10.2.1.2 The Main Statistical Methods for Developing a Scale

Item analysis, reliability analysis and factor analysis are the most commonly used statistical methods in developing the "Assessment Scale".

(1) Item analysis

Item analysis is mainly used to test the reliability of the individual components of the scale. It was adopted in the previous analysis of the "Importance Scale". The criteria of selecting data can be based on general standards, or according to the content of the scale and the purpose of evaluation. The "Importance Scale" was mainly analysed from the mean, skewness, correlation and etc. The Assessment Scale investigates whether each item has the ability to discriminate the concepts. The method is to divide the score of the item into "high score group" and "low score group" (usually the first 27% and the last 27% as the boundary), and then through the Independent Samples T-test to check whether difference in the scores of the two groups is significant. If it is significant, the item is discriminative; If not, the item itself is meaningless, since people give similar answers.

(2) Reliability analysis

Reliability analysis is mainly used to assess the reliability of the entire scale. Reliability is considered to be the accuracy of producing the same result if the measurement is re-applied (Ke Huixin et al. 2003). That is the stability and reliability of the measurement. The coefficient is adopted to reflect reliability. This book uses the Cronbach's reliability coefficient (Cronbach's α) as the reliability test index. The α is between 0–1 and the greater it is, the higher the reliability it has. What coefficient can be considered as high reliability? Different scholars have different opinions. Generally, the lowest acceptable coefficient of α is 0.65 or 0.7. If it is between 0.60 and 0.65, it should be deleted. The coefficient value between 0.7–0.8 means quite good, but the ideal value is between 0.8–0.9.

(3) Factor analysis

Factor analysis is mainly used to explore and examine the structural validity of the scale. Factor analysis includes exploratory analysis and confirmatory analysis. Exploratory factor analysis is used to explore the internal structure of credibility evaluation, and to select items for the scale. Principal components analysis and varimax rotation are used to explore the scale factor structure.

Confirmatory factor analysis is to verify the factor structure obtained by exploratory factor analysis (equivalent to a hypothetical theoretical model), and test whether the data fits the proposed factor structure. It provides more statistics for factor structure.

In this research the structural equation model (SEM) is used for confirmatory factor analysis. The SEM is commonly applied in research of other countries, but it is still at the beginning stage in China. The structural equation model is a statistical analysis technique to develop, estimate and test cause-effect relationship models. It contains a series of multivariate statistical analysis techniques such as regression analysis, factor analysis, path analysis and multivariate analysis of variance. It is a general linear statistical modelling technique that uses a theory to conduct hypothesis testing (Ke Huixin et al. 2003).

There are many ways to compare the sample's fitness. The most commonly used are:

(a) Chi-square goodness of fit (x^2 goodness-of-fit test). If the chi-square value is not significant, it is a good fit. Usually, if the ratio of the chi-square value to the degree of freedom is less than 2 or 3, the model fits good. However, the chi-square value is closely related to the sample size, so it is impossible to use the chi-square value to test the goodness of fit.
(b) Goodness of Fit Index (GFI) and Adjusted Goodness of Fit Index (AGFI). These two indexes are taken as the overall index of Mode Adequacy, and the size is between 0–1. If the value is greater than 0.9, it is regarded as the model fit observation value.
(c) Normed Fit Index (NFI), Non-Normed Fit Index (NNFI) and Comparative Fit Index (CFI). These three commonly used indexes are called Comparative Fit Indexes. They are obtained from the comparison between the fitting of the set model and the fitting of the independent model. The size is 0–1. If it is above 0.9, the model fits well.
(d) Root Mean Square Error of Approximation (RMSEA). This structural equation model fitting evaluation index has received more attention in recent years. If the RMSEA is small or equal to 0.05, it means "good" fit; 0.05 -0.08 is regarded as "relatively good" fit; 0.08 -0.10 is "moderate" fit and bigger than 0.10 means "bad" fit (Guo Zhigang 1999; Ke Huixin et al. 2003).

According to statistical requirements, the same data cannot be used for exploratory factor analysis and confirmatory factor analysis. Therefore, about 50% of the data (372 samples in total) were randomly selected from 693 samples through SPSS software for exploratory factor analysis. The other half of the data (a total of 321 samples) were used for confirmatory factor analysis, and the statistical software used was LISREL 8.20.

10.2.2 Four Selections of Indexes for the Assessment Scale

Based on the results of the "Importance Scale" three items that are not suitable for the credibility scale can be deleted. They are listed as followings.

Not reporting some conferences and government affairs that have little to with the ordinary people. (EXP24)

Not reporting negative news that may cause social unrest. (EXP15)

The media is consistent with the Party and the government under all circumstances. (EXP16)

Thus, 25 items are remained in the fourth round of selection.

10.2.2.1 Item Analysis

The item discrimination is tested. Results show that the score difference between the "high score group" and "low score group" has reached a very significant level, indicating that 25 items have good discrimination (see Table 10.1).

10.2.2.2 Exploratory Factor Analysis

(1) "Comprehensive scale" for credibility evaluation

The principal component analysis and the varimax rotation were applied on the 25 items. We obtained 4 factors through not limiting the number of factors and extracting a common factor with an eigenvalue greater than 1. The factor analysis results of the "Assessment Scale" are consistent with that of the "Importance Scale". According to the contents of the corresponding items of the factors, the four factors (F1, F2, F3, F4) are named "Social concern", "Journalistic Skills", "Media Ethics", "Journalistic Professionalism". The factor loading of each item reached 0.463, which is relatively ideal. It verified that the media credibility is composed of four dimensions. The items included in the scale is relatively comprehensive. It is also named "Comprehensive Scale for Credibility Measurement" (referred to as the Comprehensive Scale). The results of factor analysis are shown in Table 10.2 (the factor loading lower than 1 is omitted).

(2) "Basic scale" for credibility evaluation

The scale composed of 25 items is obviously not concise, and the categories of some items are not clear, where two factors' loading exceed 0.4. It is necessary to further select the items, which is on the basis of previous results that "social concern" and "journalistic professionalism" are the most important aspects for the credibility judgment of the public while "Media Ethics" and "Journalistic Professionalism" are secondary aspects. Thus the deletion of items is done according to the importance of these items.

Table 10.1 The item discrimination

Item			Number of people	Mean	Standard deviation	Standard error of mean	Statistical significance
JUD01	Factual reports	High score group	194	7.74	1.547	0.111	0.000
		Low score group	185	4.58	2.047	0.150	
JUD02	Full coverage	High score group	194	7.63	1.618	0.116	0.000
		Low score group	185	3.83	2.15	0.158	
JUD03	Complete coverage	High score group	194	7.85	1.471	0.106	0.000
		Low score group	185	4.06	2.242	0.165	
JUD04	Accurate reports	High score group	194	7.90	1.654	0.119	0.000
		Low score group	185	4.68	2.091	0.154	
JUD05	Objective without bias	High score group	194	7.55	1.557	0.112	0.000
		Low score group	185	4.28	2.084	0.153	
JUD06	Balancing without bias	High score group	194	7.76	1.467	0.105	0.000
		Low score group	185	4.56	2.116	0.156	
JUD07	Prompt and timely	High score group	194	8.42	1.383	0.099	0.000
		Low score group	185	5.59	2.163	0.159	

(continued)

Table 10.1 (continued)

Item			Number of people	Mean	Standard deviation	Standard error of mean	Statistical significance
JUD08	In-depth	High score group	194	7.96	1.382	0.099	0.000
		Low score group	185	5.04	2.127	0.156	
JUD09	Exclusive	High score group	194	7.33	1.477	0.106	0.000
		Low score group	185	4.29	2.317	0.17	
JUD10	Novel and vivid	High score group	194	7.29	1.486	0.107	0.000
		Low score group	185	4.41	2.256	0.166	
JUD11	Correcting errors	High score group	194	7.85	1.715	0.123	0.000
		Low score group	185	4.02	2.512	0.185	
JUD12	Focusing on reality	High score group	194	7.93	1.255	0.09	0.000
		Low score group	185	5.01	2.139	0.157	
JUD13	Sharp perspective	High score group	194	7.86	1.298	0.093	0.000
		Low score group	185	4.38	2.326	0.171	
JUD14	Government supervision	High score group	194	7.41	1.624	0.117	0.000
		Low score group	185	2.84	2.559	0.188	

(continued)

Table 10.1 (continued)

Item			Number of people	Mean	Standard deviation	Standard error of mean	Statistical significance
JUD18	Respect for privacy	High score group	194	8.02	1.522	0.109	0.000
		Low score group	185	4.77	2.299	0.169	
JUD19	Exposing wrongdoing	High score group	194	7.71	1.443	0.104	0.000
		Low score group	185	3.88	2.281	0.168	
JUD20	Public view	High score group	194	8.13	1.403	0.101	0.000
		Low score group	185	4.26	2.306	0.17	
JUD21	Concerned about the disadvantaged groups	High score group	194	8.33	1.36	0.098	0.000
		Low score group	185	4.56	2.243	0.165	
JUD22	Equal manner	High score group	194	8.04	1.591	0.114	0.000
		Low score group	185	4.55	2.293	0.169	
JUD23	Not high-flown	High score group	194	8.02	1.475	0.106	0.000
		Low score group	185	3.76	2.286	0.168	
JUD25	Controlling proportion of advertising	High score group	194	6.84	1.778	0.128	0.000
		Low score group	185	3.52	2.465	0.181	

(continued)

Table 10.1 (continued)

Item			Number of people	Mean	Standard deviation	Standard error of mean	Statistical significance
JUD26	No false advertising	High score group	194	7.09	2.005	0.144	0.000
		Low score group	185	2.76	2.287	0.168	
JUD27	No paid news	High score group	194	6.88	1.747	0.125	0.000
		Low score group	185	3.02	2.281	0.168	
JUD28	Elegant style	High score group	194	7.47	1.6	0.115	0.000
		Low score group	185	3.78	2.356	0.173	
JUD29	Independent editing	High score group	194	7.43	1.464	0.105	0.000
		Low score group	185	3.29	2.182	0.16	

The two items, "able to independently review and supervise government policies and actions" and "the editorial is independent from marketing so reports are not affected by money", are more related to the media system while the other aspects are directly relevant to the people. Although in the previous interviews, some have begun to focus on the media system to evaluate the media credibility, most of Chinese people make intuitive judgment of the media credibility through their "appearance". To maintain the conceptual unity in the scale, we decide to delete these two items.

"Respect for privacy" is generally considered in the credibility scale for Western countries. In a fully marketed media environment, media often strives to dig up some eye-catching news to gain profit. That news related to privacy violates ethics, and causes complaints, even though it meets the professional standards, such as "truthfulness" and "comprehensiveness". It can take "the death of the Princess Diana in a car crash chased by paparazzi" as example.

In China, the management of media is strict, and the media are not highly commercial. Privacy violation is not a prominent issue. From the results, this item's factor

Table 10.2 Factor analysis results of "comprehensive scale"

Item	F1 Social concern	F2 Journalistic skills	F3 Media ethics	F4 Journalistic professionalism
JUD19 With the courage to expose wrongdoing and high proportion of critical reports	0.735	0.156	0.184	0.275
JUD20 Concerned about the public interest in their view	0.731	0.296	0.226	0.245
JUD21 Concerned about the disadvantaged groups, including peasants, the unemployed, and the disabled	0.729	0.262	0.21	0.172
JUD22 The media treat the readers/audiences in an equal manner, instead of being arrogant	0.575	0.295	0.379	
JUD23 The news reports tell the truth, not using high-flown rhetoric	0.568	0.306	0.293	0.344
JUD14 Able to independently review and supervise the government policies and actions	0.463	0.267	0.28	0.314
JUD18 Respect for people's privacy	0.463	0.196	0.365	0.209
JUD09 Able to have exclusive news often	0.171	0.712	0.208	
JUD08 Able to provide in-depth analysis and interpretation of complex news events	0.193	0.669	0.164	0.251
JUD12 The news covers the most realistic and urgent issues in social development	0.434	0.652		0.186
JUD07 News reports are prompt and timely	0.195	0.643		0.353
JUD10 The news layout or program is novel, vivid and unique	0.123	0.602	0.24	0.184

(continued)

Table 10.2 (continued)

Item	F1 Social concern	F2 Journalistic skills	F3 Media ethics	F4 Journalistic professionalism
JUD11 Apologise and correct wrong reports once they have been found	0.221	0.589	0.258	0.203
JUD13 Good at finding problems, and the reports have a sharp perspective	0.395	0.587	0.127	0.211
JUD26 Not posting false advertising	0.211	0.197	0.799	0.21
JUD27 Not publishing or broadcasting advertised or paid news, soft-sell advertising	0.266	0.11	0.795	0.239
JUD25 The proportion of advertising is appropriate	0.146	0.213	0.787	
JUD28 The reporting style is elegant, without vulgar news	0.22	0.182	0.72	0.289
JUD29 The editorial is independent from marketing so reports are not affected by money	0.363	0.203	0.612	0.315
JUD01 Factual reports, news reports do not contain false, speculative and fictitious elements	0.319	0.141	0.116	0.711
JUD04 The news report is particularly accurate	0.215	0.214	0.228	0.694
JUD02 Full coverage of news events, not avoiding any important facts	0.339	0.193	0.232	0.670
JUD06 For controversial news, balancing the coverage of each party without bias	0.176	0.411	0.136	0.640
JUD05 Objectively present the original appearance of the event without bias		0.436	0.306	0.598
JUD03 Report all news events that the public want to know	0.449	0.207	0.164	0.523

Note: The ratios of the explanatory variances after the rotation of the four factors are: 16.056%, 15.971%, 15.364%, 14.154%; Factor cumulative explanatory variance ratio is 61.546%

loading on the "social concern" is not high. Perhaps the public don't connect this issue to the "social concern" category, so this item is deleted.

The two items, "JUD12 the news covers the most realistic and urgent issues in social development" and "JUD13 Good at finding problems and having a sharp perspective" have a relatively high loading at two factors, "social concern" and "journalistic skills". It means two items have multiple meanings, so they are deleted.

The item, "JUD11 apologise and correct wrong reports once they have been found", is only an incident, not the normal state, so it is deleted.

The relatively low importance has been found in previous result on the item, "JUD10 the news layout or program is novel, vivid and unique", and the commonality of this item is low, so it is deleted.

In addition, two items in "journalistic professionalism", "JUD06 for controversial news, balancing the coverage of each party without bias" and "JUD05 objectively present the original appearance of the event without bias", show a high loading on the factor of "journalistic skills" (exceeds 0.4). It may be influenced by the items, "focusing on reality" and "sharp perspective" in "journalistic skills". Considering the importance of the factor, we decide to keep the two items for further analysis.

Also the loading of the item "JUD03 report all news events that the public want to know" on the "social concern" exceeds 0.4, which may be influenced by the way of expression. The original meaning of this item is the media reports the events comprehensively, without ignoring or deliberately concealing some important events. Its essential meaning is to guarantee people's right to know. In this way the item has the meaning of "protecting the public interest", so the loadings on "social concern" and "journalistic professionalism" exceed 0.4. For "Complete coverage" is an important factor affecting credibility, it is not deleted. The statement of the item "complete coverage" will be improved in the following studies.

The structure of scale has been modified after the deletion. The factor analysis is conducted on the remaining 18 items. The principal components analysis and the varimax rotation were applied on the 18 items. We obtained 4 factors through not limiting the number of factors and extracting a common factor with an eigenvalue greater than 1. In the simplified scale, item has a clearer attribution. The contribution rate of factor explanatory variance reaches 67.5%. In general, the structure of the scale is relatively clear and meet the theoretical assumption, so it can be used as a basic scale for the further related research. It is named as "Basic Scale for Credibility Measurement", referred to as the basic scale (see Table 10.3).

(3) "Concise Scale"

We make a further attempt to further simplify and optimize the scale.

The item, "JUD03 Report all news events that the public want to know", shows a loading more than 0.4 on the factor of "social concern", indicating that the item has ambiguous meaning which has been analysed in previous section. We may try to improve it in following study, but it is deleted in the Concise Scale.

There are some similarities and repetitions in the meanings of the two items ("JUD23 The news reports tell the truth, not using high-flown rhetoric.", and "JUD22 The media treat the readers/audiences in an equal manner, instead of being arrogant").

Table 10.3 Factor analysis results of "basic scale"

Item	F1 Journalistic professionalism	F2 Social concern	F3 Media ethics	F4 Journalistic skills
JUD01 Factual reports, news reports do not contain false, speculative and fictitious elements	0.729	0.314	0.146	0.102
JUD04 The news report is particularly accurate	0.710	0.177	0.24	0.209
JUD02 Full coverage of news events, not avoiding any important facts	0.699	0.341	0.231	0.119
JUD06 For controversial news, balancing the coverage of each party without bias	0.630	0.162	0.137	0.442
JUD05 Objectively present the original appearance of the event without bias	0.587		0.309	0.476
JUD03 Report all news events that the public want to know	0.576	0.438	0.15	0.13
JUD20 Concerned about the public interest in their view	0.281	0.762	0.211	0.254
JUD21 Concerned about the disadvantaged groups, such as rural people, the unemployed, and the disabled	0.187	0.740	0.197	0.274
JUD19 With the courage to expose wrongdoing and high proportion of critical reports	0.341	0.723	0.183	

(continued)

Table 10.3 (continued)

Item	F1 Journalistic professionalism	F2 Social concern	F3 Media ethics	F4 Journalistic skills
JUD22 The media treat the readers/audiences in an equal manner, instead of being arrogant	0.372	0.607	0.359	0.306
JUD23 The news reports tell the truth, not using high-flown rhetoric		0.579	0.297	0.265
JUD25 The proportion of advertising is appropriate		0.167	0.824	0.127
JUD26 Not posting false advertising	0.233	0.22	0.823	0.153
JUD27 Not publishing or broadcasting advertised or paid news, soft-sell advertising	0.257	0.269	0.789	
JUD28 The reporting style is elegant, without vulgar news	0.299	0.217	0.698	0.195
JUD07 News reports are prompt and timely	0.327	0.208		0.739
JUD08 Able to provide in-depth analysis and interpretation of complex news events	0.236	0.225	0.16	0.722
JUD09 Able to have exclusive news often		0.276	0.198	0.708

Note: The ratios of the explanatory variances after the rotation of the four factors are: 18.634%, 17.688%, 17.433%, 13.727%; factor cumulative explanatory variance ratio is 67.482%

From analysis of the factor loading, the meaning of item JUD23 is not very clear, and it has a certain relationship with the other three factors. It is deleted for simplification of the scale.

From the previous analysis, we find that most people in China have a weak ability to recognize paid news, and the importance of this item is not high. Therefore, this item can be deleted.

In addition, similar to the previous analysis, in "professionalism", the two items "balancing without bias" and "objective without bias" have a more obvious relationship with the "journalistic skills", and the factor loading is higher than 0.4. The item "prompt and timely" in "journalistic skills" show a high factor loading on "professionalism" (0.327). Previous results show the importance of "in-depth" and "exclusive news" is low, and the group "lower than a bachelor's degree" paid less attention to it. Considering the importance of the factor "journalistic skills" is the lowest among four factors, we can delete this factor to avoid ambiguity. However, the importance of the "timely" is quite high, so it is remained in the "professionalism" factor.

Then, the factor analysis was conducted again on the remaining 13 factors. The principal component analysis and the varimax rotation were still applied. We obtained 3 factors through not limiting the number of factors and extracting a common factor with an eigenvalue greater than 1. The results show that the factor loading is above 0.6, and there is no case that the loading of an item on both factors exceeds 0.4. The remaining three factors represent the three dimensions of media credibility, which are "Social concern", "Media Ethics" and "Journalistic Professionalism". The content is ideal, so the scale is named "Concise Scale of Credibility Measurement" (referred to as the "Concise Scale"). The results of factor analysis are shown in Table 10.4.

10.2.3 Reliability and Validity Test of the Three Scales

Through the above exploratory factor analysis, the three different scales are obtained, which are the "Comprehensive Scale" with 25 items and 4 dimensions, "Basic Scale" with 18 indexes and 4 dimensions and "Concise Scale" with 13 indexes and 3 dimensions.

The reliability and validity of these three scales are tested below. The reliability coefficients Cronbach's alpha is the main indicator to test the internal consistency of the whole scale and each constituent dimension (as a subscale). The structural equation model was used for confirmatory factor analysis of the scale structure to test the structural validity of the scale. The reliability test used all 692 samples, and confirmatory factor analysis used 321 samples that were not adopted in exploratory factor analysis.

10.2.3.1 Reliability and Structure Validity Test of the "Comprehensive Scale"

(1) Reliability test

The overall reliability coefficient alpha of the scale is 0.9474, which is a fairly high. The deletion of any item will cause the decrease of the reliability coefficient. The coefficient alpha of the four subscales has reached a high level: "Journalistic

Table 10.4 Factor analysis results of "concise scale"

Item	F1 Journalistic professionalism	F2 Social concern	F3 Media ethics
JUD06 For controversial news, balancing the coverage of each party without bias	0.644	0.361	0.15
JUD05 Objectively present the original appearance of the event without bias	0.625	0.360	0.245
JUD04 The news report is particularly accurate	0.707	0.223	0.224
JUD01 Factual reports, news reports do not contain false, speculative and fictitious elements	0.759		0.299
JUD07 News reports are prompt and timely	0.784	0.237	0.139
JUD02 Full coverage of news events, not avoiding any important facts	0.644	0.299	
JUD20 Concerned about the public interest in their view	0.261	0.754	0.176
JUD21 Concerned about the disadvantaged groups, such as peasants, the unemployed, and the disabled	0.302	0.814	0.222
JUD19 With the courage to expose wrongdoing and high proportion of critical reports	0.259	0.792	0.171
JUD22 The media treat the readers/audiences in an equal manner, instead of being arrogant	0.228	0.601	0.370
JUD25 The proportion of advertising is appropriate		0.186	0.858
JUD26 Not posting false advertising	0.254	0.262	0.826
JUD28 The reporting style is elegant, without vulgar news	0.345	0.228	0.704

Note: The ratios of the explanatory variances after the rotation of the three factors are: 26.019%, 21.778%, 18.492%; Factor cumulative explanatory variance ratio is 66.289%

Professionalism" subscale, α = 0.8790; "Social Concern subscale", α = 0.8890; "Journalistic Skills" subscale, α = 0.8536; "Media Ethics", α = 0.8723. The results of the reliability analysis of the "Comprehensive Scale" is shown in Table 10.5 (the specific data of reliability analysis of each subscale is omitted).

(2) Structural validity test

Table 10.5 Reliability test data of "comprehensive scale"

Item	Scale mean if item deleted	Scale variance if item deleted	Corrected item-total correlation	Alpha if item deleted
JUD01 Factual reports	143.4921	11 310.4439	0.6427	0.9453
JUD02 Full coverage	143.9481	11 291.7545	0.6899	0.9447
JUD03 Complete coverage	143.6219	11 294.9349	0.6661	0.9449
JUD04 Accurate reports	143.4560	11 308.3756	0.6305	0.9454
JUD05 Objective without bias	143.7258	11 309.4825	0.6288	0.9454
JUD06 Balancing without bias	143.4589	11 308.9799	0.6424	0.9453
JUD07 Prompt and timely	142.5224	11 320.0736	0.6036	0.9457
JUD08 In-depth	143.0967	11 315.1597	0.6138	0.9456
JUD09 Exclusive	143.8398	11 321.3890	0.5210	0.9466
JUD10 Novel and vivid	143.7951	11 320.2383	0.5408	0.9464
JUD11 Correcting errors	143.6537	11 295.7065	0.6106	0.9456
JUD12 Focusing on reality	143.1328	11 318.2974	0.6142	0.9456
JUD13 Sharp perspective	143.3954	11 305.9793	0.6407	0.9453
JUD14 Government supervision	144.4603	11 276.4395	0.6529	0.9452
JUD18 Respect for privacy	143.1847	11 312.2924	0.5893	0.9458
JUD19 Exposing wrongdoing	143.7994	11 291.1837	0.6864	0.9447
JUD20 Public view	143.4199	11 287.9925	0.7284	0.9442
JUD21 Concerned about the disadvantaged groups	143.2078	11 297.1186	0.6681	0.9449
JUD22 Equal manner	143.3636	11 298.4890	0.6641	0.9450
JUD23 Not high-flown	143.7994	11 279.3051	0.7259	0.9442

(continued)

Table 10.5 (continued)

Item	Scale mean if item deleted	Scale variance if item deleted	Corrected item-total correlation	Alpha if item deleted
JUD25 Controlling proportion of advertising	144.5743	11 312.2159	0.4972	0.9471
JUD26 No false advertising	144.7071	11 278.5600	0.6511	0.9452
JUD27 No paid news	144.6291	11 292.8782	0.6190	0.9455
JUD28 Elegant style	143.9784	11 299.8738	0.5986	0.9458
JUD29 Independent editing	144.1876	11 286.4648	0.6704	0.9449

Note: Alpha = 0.9474

The structural equation modelling technique was adopted to make the confirmatory factor analysis of the factor structure of "Comprehensive Scale". The result is $x^2 =$ 920.94, $df = 269$, $P = 0.0$, the ratio of chi-square value to degree of freedom is more than 3, $P < 0.05$ (indicating the model has no fitted data); GFI = 0.81, AGFI = 0.77, the goodness-of-fit index does not reach 0.9 (indicating that the model has no fitted data); NFI = 0.81, NNFI = 0.84, CFI = 0.86, comparative fit index does not reach 0.9 (indicating that the model has no fitted data); RMSEA = 0.087; the root mean square of the approximate error is between 0.8 -0.1, a moderate fit of the model. Overall, the model fit of the scale is not ideal. The factor structure model and path coefficients of the "Comprehensive Scale" are shown in Fig. 10.1, where 25 observed variables are 25 scale indexes, and 4 latent variables are 4 corresponding factors.

10.2.3.2 The Reliability and Structural Validity of the "Basic Scale"

(1) Reliability test

The overall reliability coefficient alpha of the scale is 0.9308, which is quite high. The deletion of any item will cause the decrease of the reliability coefficient. The coefficient alpha of the four subscales are relatively high, except that of "Journalistic Skills", Journalistic Professionalism subscale, $\alpha = 0.8790$; Social Concern subscale, $\alpha = 0.8820$; Journalistic Skills subscale, $\alpha = 0.7440$; Media Ethics subscale, $\alpha =$ 0.8790. The results of the reliability analysis of the "Basic Scale" is shown in Table 10.6 (the specific data of reliability analysis of each subscale is omitted).

(2) Structural validity test

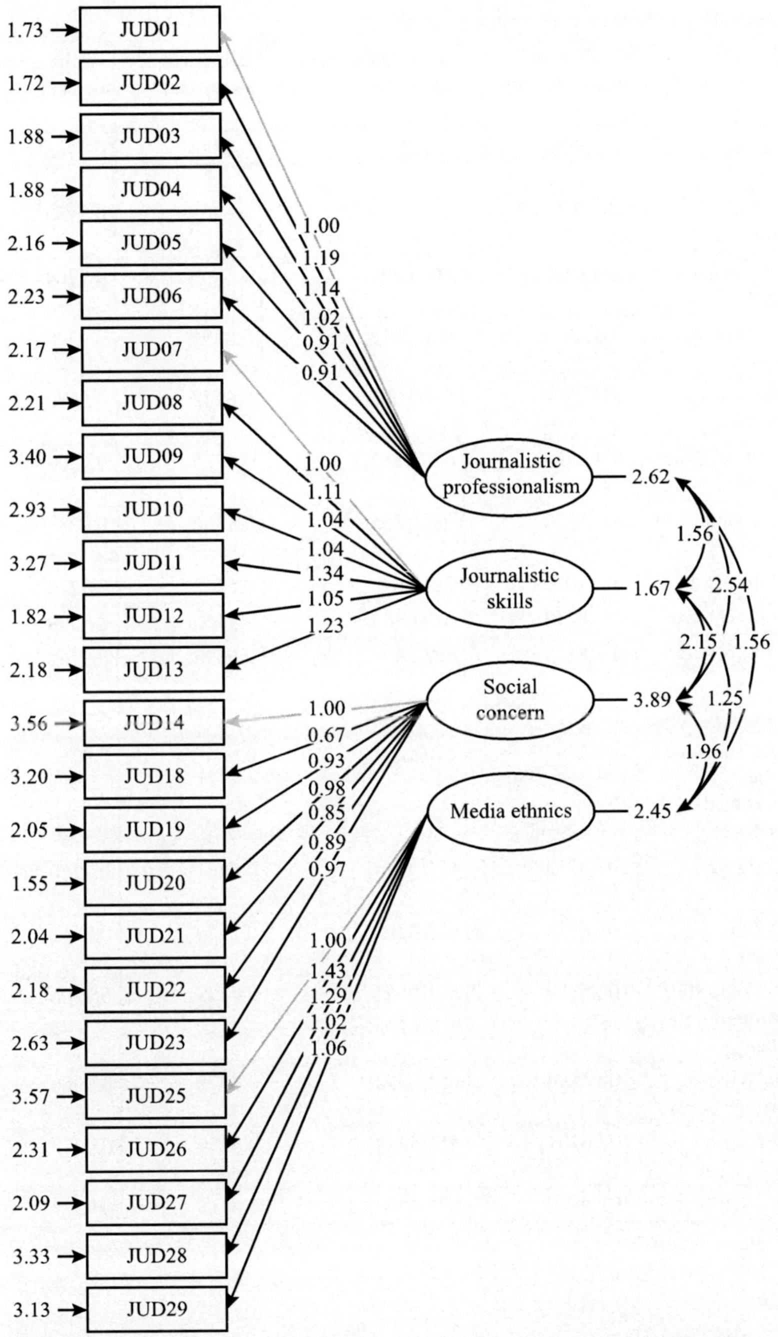

Fig. 10.1 The factor structure model and path coefficients of the "comprehensive scale"

Table 10.6 Reliability test data of "basic scale"

Item	Scale mean if item deleted	Scale variance if item deleted	Corrected item-total correlation	Alpha if item deleted
JUD01 Factual reports	101.5036	678.2648	0.6534	0.9266
JUD02 Full coverage	101.9596	665.0822	0.6967	0.9255
JUD03 Complete coverage	101.6335	669.0880	0.6569	0.9264
JUD04 Accurate reports	101.4675	676.8013	0.6399	0.9269
JUD05 Objective without bias	101.7374	679.0003	0.6248	0.9272
JUD06 Balancing without bias	101.4704	678.2582	0.6425	0.9268
JUD07 Prompt and timely	100.5339	688.1365	0.5846	0.928
JUD08 In-depth	101.1082	684.5678	0.5953	0.9278
JUD09 Exclusive	101.8514	689.6383	0.4966	0.9299
JUD19 Exposing wrongdoing	101.811	667.246	0.6698	0.9261
JUD20 Public view	101.4315	663.6936	0.7241	0.925
JUD21 Concerned about the disadvantaged groups	101.2193	670.1223	0.664	0.9263
JUD22 Equal manner	101.3752	672.4024	0.648	0.9266
JUD23 Not high-flown	101.811	658.4396	0.7124	0.9251
JUD25 Controlling proportion of advertising	102.5859	681.0551	0.4889	0.9306
JUD26 No false advertising	102.7186	655.2372	0.6573	0.9266
JUD27 No paid news	102.6407	665.5282	0.6259	0.9272
JUD28 Elegant style	101.9899	671.0909	0.6013	0.9277

Note: Alpha = 0.9308

The structural equation modelling technique was adopted to conduct the confirmatory factor analysis of the factor structure of the "Basic Scale". The result is $x^2 = 421.06$, $df = 129$, $P = 0.0$, the ratio of chi-square value to degree of freedom is more than 3, $P < 0.05$ (indicating the model has no fitted data, but it is very close to 0.9); GFI = 0.87, AGFI = 0.83, the goodness-of-fit index does not reach 0.9 (indicating that the model has no fitted data, but it is very close to 0.9); NFI = 0.87, NNFI = 0.89, CFI = 0.91, two of the comparative fit indexes do not reach 0.9 (indicating that it is moderately fitted); RMSEA = 0.084; the root mean square of the approximate error is between 0.8–0.1, a moderate fit of the model. Overall, the model fit of the scale has been improved compared to the "comprehensive scale", but it is still not ideal. The factor structure model and path coefficients of the "Basic Scale" are shown in Fig. 10.2, where 18 observed variables are 18 scale indexes, and 4 latent variables are 4 corresponding factors.

10.2.3.3 The Reliability and Structural Validity of the "Concise Scale"

(1) Reliability test

The overall reliability coefficient alpha of the scale is 0.9079, which is quite high. The deletion of any item will cause the decrease of the reliability coefficient. The coefficient alpha of the three subscales are relatively high: Journalistic Professionalism subscale, $\alpha = 0.8627$; Social Concern subscale, $\alpha = 0.8588$; Media Ethics subscale, $\alpha = 0.7881$. The results of the reliability analysis of the "Concise Scale" is shown in Table 10.7 (the specific results of the reliability analysis of each subscale are omitted).

(2) Structural validity test

The structural equation modelling technique was adopted to conduct the confirmatory factor analysis of the factor structure of "Concise Scale". The result is $x^2 = 207.10$, df = 62, P = 0.0, the ratio of chi-square value to degree of freedom is more than 3, $P < 0.05$ (indicating the model has no fitted data); GFI = 0.91, AGFI = 0.87, the goodness-of-fit index is higher or close to 0.9 (indicating that the model has fitted data); NFI = 0.90, NNFI = 0.91, CFI = 0.93, comparative fit indexes reach 0.9 (indicating the model fits data); RMSEA = 0.086; the root mean square of the approximate error is between 0.8 -0.1, a moderate fit of the model. Overall, the model fitting situation of the scale has been further improved, which can be regarded as a good fit. The factor structure model and path coefficients of the "Concise scale" are shown in Fig. 10.3, in which 13 observation variables are 13 scale indexes and 3 latent variables are 3 corresponding factors.

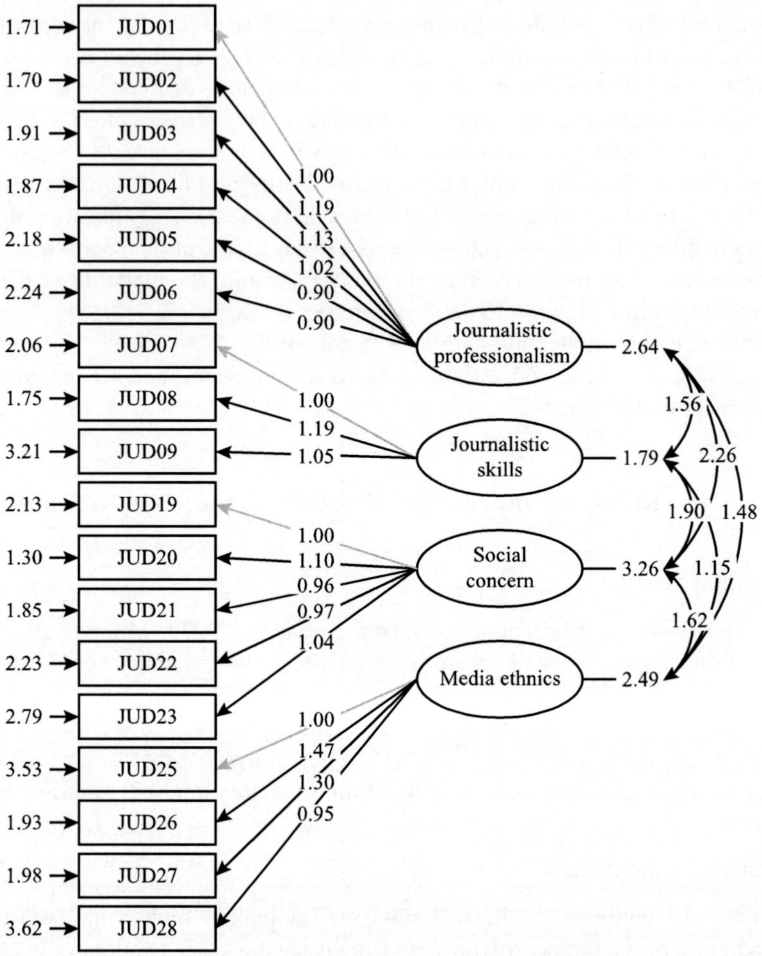

Fig. 10.2 The factor structure model and path coefficients of the "basic scale"

10.3 Summary

In this chapter the credibility evaluation indexes are further selected through item analysis, reliability analysis, exploratory and confirmatory factor analysis. The reliability and validity of the results are tested. The three scales are used to briefly describe and analyse the credibility evaluation status.

The optimal scale can balance accuracy and conciseness, so the scale is simplified while maintain their accuracy. Since the research of media credibility is still at early stage, the research does not simplify the scale too much. It will be improved and will be concise after nationally sampling.

Table 10.7 Reliability test data of "concise scale"

Item	Scale mean if item deleted	Scale variance if item deleted	Corrected item-total correlation	Alpha if item deleted
JUD01 Factual reports	72.0635	347.6145	0.6596	0.8997
JUD02 Full coverage	72.5195	339.1141	0.6894	0.8981
JUD04 Accurate reports	72.0274	346.5527	0.6451	0.9001
JUD05 Objective without bias	72.2973	349.2352	0.6147	0.9014
JUD06 Balancing without bias	72.0303	347.3705	0.6514	0.9000
JUD07 Prompt and timely	71.0938	357.4262	0.551	0.9039
JUD19 Exposing wrongdoing	72.3709	341.3550	0.6529	0.8997
JUD20 Public view	71.9913	337.8178	0.7217	0.8968
JUD21 Concerned about the disadvantaged groups	71.7792	342.7041	0.6564	0.8996
JUD22 Equal manner	71.9351	345.1822	0.6290	0.9007
JUD25 Controlling Proportion of advertising	73.1457	350.9426	0.4717	0.9083
JUD26 No false advertising	73.2785	333.5567	0.6307	0.9011
JUD28 Elegant style	72.5498	344.0427	0.5837	0.9028

Note: Alpha = 0.9079

The development of a perfect Assessment Scale requires continuous theoretical and empirical verification. Although the scale indexes have undergone multiple rounds of selection and testing, further testing is still vital. Therefore, this research couldn't obtain the "best" scale, but proposed the three scales with different levels of conciseness.

The three scales have reached acceptable or ideal levels in terms of reliability, surface validity, and structural validity. In the following study different scales can be applied according to the situation. In a special study of the media credibility assessment scale it is recommended to use the "Comprehensive Scale" and indexes can be re-selected. If the media credibility is only a variable in study, a more simplified scale can be used.

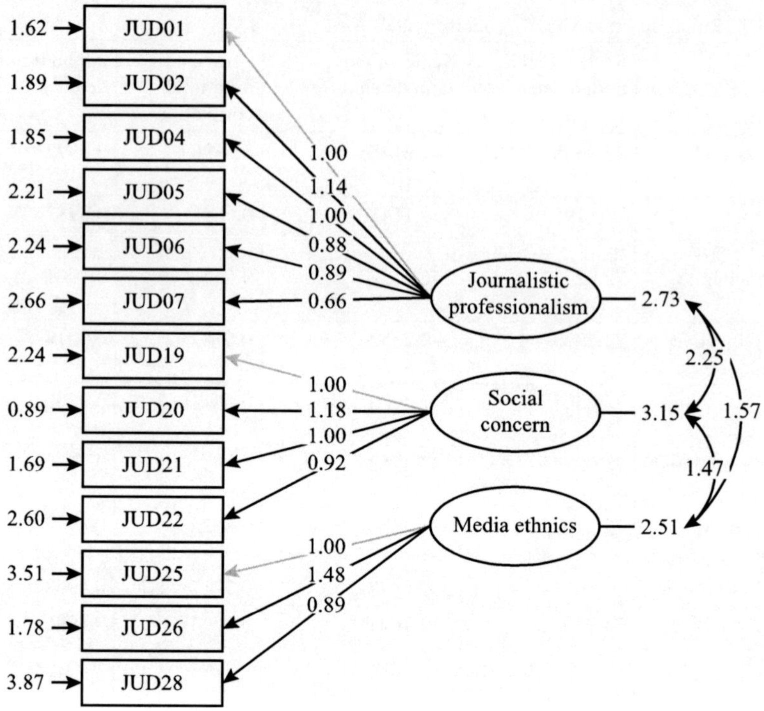

Fig. 10.3 The factor structure model and path coefficients of the "Concise Scale"

Chapter 11
Media Credibility Assessment Based on the Survey Specimen

It is pointed out that "China's mass media" has different meanings in Chap. 1 that explains the research concept. It can be a macro media in an abstract sense, or the media with various forms (such as TV, newspapers, radio, magazines, Internet, etc.), or specific kinds of media (such as CCTV, *People's Daily* or *Caijing Magazine*, etc.), or different types of news (such as legal news, public safety news, financial news, etc.), or a specific program or news report (such as *Focus Interview*, a news report on the Iraq War of the *Global Times*, etc.). Therefore, the subjects of media credibility assessment are diverse.

The key to evaluate media credibility is to adopt a proper method. The methods have been introduced and evaluated in literature review. The most commonly used are multi-dimensional assessment scale, rating, selecting the relative credible assessment of "most trusted media", etc. The assessment methods are determined by a variety of factors such as evaluation subjects, research objectives and research funding support.

The assessment scale developed through theoretical and empirical data testing is a multi-dimensional tool for measuring the China's macro media credibility. This survey used several other commonly methods and assessed the credibility of different media. This chapter introduces and analyses the evaluation results.

This study, as an exploratory research before the nationwide survey, is to initially construct the assessment scale. The 692 samples were obtained by non-random sampling in Beijing. The results obtained from the samples cannot display the whole situation of China, but it can roughly reflect the current situation of China's media credibility. Moreover, if the three scales proposed previously and other commonly used scales are applied at the same time, the differences will be noticed, which is an important part of the study of credibility assessment. The following is objective description of the evaluation results, and a comparative analysis of the assessment methods and the credibility evaluation of groups with different education backgrounds. Issues related to the overall status of the China's media credibility will not be discussed further.

G. Yu, *Research on the Communication Effects and Mass Media Credibility in China*,
https://doi.org/10.1007/978-981-19-6242-4_11

11.1 The Overall Credibility of China's Mass Media

11.1.1 *Scores of Indexes in Media Credibility Evaluation*

In this survey, 25 indexes were used to measure the credibility of the macro media, which represent different media performance. The score ranges from 0 to 10 points, with 0-point indicating "have done very poorly in this aspect", and 10 points indicating "have done very well in this aspect"; 6 points refer to pass. The results of indexes are shown in Table 11.1 (arranging in descending order according to the mean score).

The results show that, except for "News reports are prompt and timely" with the score higher than 7, the other items' scores are relatively low, around the 6-point pass line.

11.1.2 *Overall Evaluation Results of Media Credibility*

The three scales, "Comprehensive Scale", "Basic Scale" and "Concise Scale" can provide evaluation of the credibility of the macro media, which contain 25, 18 and 13 indexes respectively. Results vary with indexes of scale. Chapter 9 obtained the weighted reference values of indexes based on the "Importance Scale". After weighting, the credibility assessment results will change. Table 11.2 lists the overall evaluation results of media credibility obtained from different scales and weighting.

The results show that the credibility evaluation obtained from the "Concise Scale" is slightly higher than the other two scales, and the scores of the "Comprehensive Scale" are close to that of the "Basic Scale".

After weighting, the evaluation results of the "Comprehensive Scale" and the "Basic Scale" rose slightly, from lower than 6 points to pass. The reason for the increase after weighting is that the indexes with low evaluation scores have low weight.

Media credibility evaluation results are based on a sample in which the majority has a better education backgrounds and be at young or middle age. Previous studies have found education levels have a significant correlation with credibility evaluation, and that people with a higher education level gave a lower evaluation of media credibility. For all Chinese people, the credibility score should be higher than the results in Tables 11.1 and 11.2. The following is a comparative analysis of the evaluation of groups with different education backgrounds.

Table 11.1 Scores of indexes of media credibility

Ranking	Index	Sample size	Mean
1	News reports are prompt and timely	692	7.159
2	Able to provide in-depth analysis and interpretation of complex news events	692	6.584
3	The news covers the most realistic and urgent issues in social development	692	6.549
4	Respect for people's privacy	692	6.503
5	Concerned about the disadvantaged groups, including peasants, the unemployed, and the disabled	692	6.473
6	The media treat the readers/audiences in an equal manner, instead of being arrogant	692	6.318
7	Good at finding problems, and the reports have a sharp perspective	692	6.286
8	Concerned about the public interest in their view	692	6.263
9	The news report is particularly accurate	692	6.227
10	For controversial news, balancing the coverage of party without bias.	692	6.223
11	Factual reports, news reports do not contain false, speculative and fictitious elements	692	6.189
12	Report all news events that the public want to know	692	6.059
13	Apologise and correct wrong reports once they have been found	692	6.026
14	Objectively present the original appearance of the event without bias	692	5.955
15	The news layout or program is novel, vivid and unique	692	5.887
16	The news reports tell the truth, not using high-flown rhetoric	692	5.883
17	With the courage to expose wrongdoing and high proportion of critical reports	692	5.880
18	Able to have exclusive news often	692	5.841
19	Full coverage of news events, not avoiding any important facts	692	5.736
20	The reporting style is elegant, without vulgar news	692	5.702
21	The editorial is independent from marketing, so reports are not affected by money	692	5.491
22	Able to independently review and supervise government policies and actions	692	5.218
23	The proportion of advertising is appropriate	692	5.110
24	Not publishing or broadcasting advertised or paid news, soft-sell advertising	692	5.056
25	Not posting false advertising	692	4.974

Table 11.2 Overall evaluation results of credibility

Scale	Sample size	Overall score of the media
"Comprehensive Scale" (25 indexes, unweighted)	692	5.984
"Comprehensive Scale" (25 indexes, weighted)	692	6.009
"Basic Scale" (18 indexes, unweighted)	692	5.980
"Basic Scale" (18 indexes, weighted)	692	6.010
"Concise Scale" (13 indexes, unweighted)	692	6.482
"Concise Scale" (13 indexes, weighted)	692	6.302

11.1.3 Evaluation Results by Groups with Different Education Backgrounds and Comparison of Differences Among Groups

Previous studies show that education backgrounds have a significant impact on media credibility evaluation, so it is necessary to describe individual group, including "overall sample", "with a bachelor's degree or higher", "undergraduate" and "lower than a bachelor's degree". The differences among groups are analysed, shown in Table 11.3.

A one-way analysis of variance was performed on the mean scores of the three groups on the 25 indexes. The results show that the score difference among three groups on the 25 indexes reached a significant level, indicating that the evaluation of media is correlated to one's education backgrounds and life experience.

The multiple comparison results of the differences among the three groups reflect that the significant differences are mainly between the group "with a bachelor's degree or higher", and "lower than a bachelor's degree". The evaluation by those with lower education level is generally higher than those with higher education level. The most demanding group is the one "with a bachelor's degree or higher" and more life experience. Their evaluation scores of 21 indexes among the total 25 are lower than pass.

11.2 Assessment of Different Media and News Credibility

11.2.1 The Credibility Evaluation Results of Different Media

This survey compares the credibility of five media, including radio, magazine, newspaper, television, and news website. The method is to choose "the most trusted one among several types of media". It is the most commonly used by Roper, and is called "relative credibility evaluation". It has been found inadequacy at some aspects, like accuracy. However, this method is a simple and valid one to observe change of media with emergency of new media and increase of its credibility. The result obtained by

Table 11.3 The evaluation scores and differences among groups with different education backgrounds

Item	Undergraduates	Lower than a bachelor's degree	With a bachelor's degree or higher	Overall sample	Significance (P)
Prompt and timely	7.408	7.253	6.846	7.159	0.005
In-depth	6.768	6.687	6.324	6.584	0.037
Focusing on reality	6.719	6.871	6.109	6.549	0.000
Respect for privacy	6.412	7.041	6.113	6.503	0.000
Concerned about the disadvantaged groups	6.737	7.023	5.745	6.473	0.000
Equal manner	6.539	6.811	5.680	6.318	0.000
Sharp perspective	6.456	6.677	5.785	6.286	0.000
Public view	6.417	6.848	5.607	6.263	0.000
Accurate reports	6.149	6.687	5.895	6.227	0.000
Balancing without bias	6.088	6.885	5.765	6.223	0.000
Factual reports	6.127	6.641	5.850	6.189	0.000
Complete coverage	5.846	6.862	5.551	6.059	0.000
Correcting errors	5.991	6.687	5.478	6.026	0.000
Objective without bias	6.079	6.203	5.623	5.955	0.006
Novel and vivid	5.882	6.419	5.425	5.887	0.000
Not high-flown	5.868	6.728	5.154	5.883	0.000
Exposing wrongdoing	5.912	6.650	5.174	5.880	0.000
Exclusive	6.000	6.184	5.393	5.841	0.000
Full coverage	5.575	6.525	5.190	5.736	0.000
Elegant style	5.478	6.424	5.275	5.702	0.000
Independent editing	5.447	6.207	4.903	5.491	0.000
Government supervision	5.127	6.120	4.510	5.218	0.000

(continued)

Table 11.3 (continued)

Item	Undergraduates	Lower than a bachelor's degree	With a bachelor's degree or higher	Overall sample	Significance (P)
Proportion of advertising	5.127	5.562	4.696	5.110	0.001
No paid news	4.912	5.728	4.599	5.056	0.000
No false advertising	4.864	5.724	4.417	4.974	0.000
Overall mean	5.997	6.538	5.484	5.984	0.000

this method is the rate of being the "most trusted person", and the result is presented in the form of percentage.

The question has referred to the methods of researchers in other countries with the scenario that "inconsistent reporting of the same event". If there is a dispute about a report, people will rely on the media they trust. Results show people's reliance on media, which reflect media credibility. The question is as followings:

> *If you find that the following five types of media report the same event differently,*
> *Which one will you trust?*
> 1. *Broadcast* 2. *Magazine* 3. *Newspaper* 4. *TV* 5. *News website*

The results of the overall sample and the three groups are shown in Table 11.4 (the items are arranged in descending order according to the results of the overall sample).

From the overall sample data, a considerable number of people chose TV and websites as the most trusted, with the proportions of 44.4% and 31.2% respectively. The proportion of people who choose newspapers as the most reliable media is 17%; The proportions of magazine and radio are relatively low, less than 5%. Since television gained the popularity in the 1960s and 1970s, most studies have found that

Table 11.4 Results of the "relative credibility" of the five media Unit: %

Item	Undergraduates	Lower than a bachelor's degree	With a bachelor's degree or higher	Overall
TV	45.1	57.1	32.5	44.4
News website	29.5	17.0	45.4	31.2
Newspaper	18.8	14.6	17.5	17.0
magazine	4.5	8.5	2.9	4.4
radio	2.2	2.8	1.7	3.0
Sample size	224	212	240	676

TV has become the most trusted medium. The overall results show that it is still the most trusted one.

Comparing the three groups with different education backgrounds, people lower than a bachelor's degree have a higher dependence on TV and less trust on newspapers and websites. An interesting phenomenon is that the group with a bachelor's degree or higher trust websites most (45.4%) when it compares to other groups, and the proportion of TV is only 32.5%.

Even the "undergraduate" and group "lower than a bachelor's degree" trust websites more than newspapers. The results of the "relative credibility" are closely related to the popularity of the type of media. Most of the respondents are at young or middle-aged, so the proportion of the samples using online media should be higher than the whole situation. The results of this survey reveal a new trend that the highly educated, middle-aged or young people have begun to rely on the Internet more than the traditional media—TV.

The websites in China generally are not entitled to interview, and their source are mainly from other traditional media. Their gaining trust is probably related to the openness and interactivity of online forums, re-processing of information (for example, news websites often list views of both sides, other relevant events and background information through links), as well as other factors such as the access to foreign news websites.

11.2.2 Results of Credibility Evaluation of Different Types of News

For the credibility evaluation of different types of news reports, this survey adopted a scoring system with 0–10 points. A minimum of 0 point means complete distrust, a maximum of 10 points means complete trust, and 6 points refer to pass. Two questions were designed as followings;

In general, how much do you believe those types of reports on the media? Please score the degree of trust for each item.

1. *Domestic politics*	( )	2. *World news*	( )
3.*Business*	( )	4. *Life*	( )
5. *Social news*	( )	6. *Entertainment*	( )
7. *Sports*	( )	8. *Special Campaign*	( )
9. *Legal news*	( )		

In general, do you trust the following sensitive reports? Please score the degree of trust for each item.

1. *Fighting corruption*	( )	2. *unemployment*	( )
3. *Disaster and accident*	( )	4. *Public safety*	( )

Table 11.5 Trust scores of several major types of news

News type	Sample size	Minimum	Maximum	Mean	Standard deviation
Sports	625	0	10	8.11	1.68
World	633	0	10	7.63	1.81
Life	579	0	10	7.36	1.90
Legal	609	0	10	7.27	1.96
Business	603	0	10	7.17	1.78
Social	610	0	10	7.04	1.88
Domestic politics	623	0	10	6.86	2.19
Entertainment	621	0	10	6.19	2.36
Special Campaign	565	0	10	5.84	2.49

Table 11.6 Trust scores of special news

News type	Sample size	Minimum	Maximum	Mean	Standard deviation
Disaster and accidents	673	0	10	6.91	2.40
Public safety	663	0	10	6.39	2.31
Unemployment	667	0	10	6.10	2.33
Fighting corruption	673	0	10	5.92	2.61

The results are shown in Tables 11.5 and 11.6 (the items are arranged in descending order of the average score).

In Tables 11.5 and 11.6, the types of news scoring above 7 are the news with less relation to politics. Except for the "entertainment" and "special campaign", it can be found that the news' relation to the politics is inversely proportional to the news credibility.

While entertainment news has little relation to politics, it has a low score. What's more, the scores of "special campaign", "unemployment and "fighting corruption" are even lower than those of entertainment.

One-way analysis of variance was used to compare the scores of the groups with different education backgrounds. Results are shown in Table 11.7 (items are arranged in descending order of the mean).

The results show the difference of evaluation scores among three groups for "sports", "world" and "life" is not significant at the 95% confidence level. The significance of the "entertainment" is low at the 99% confidence level. Others are significant at a 99% confidence level. Through multiple comparisons of the scoring results among groups, the differences can be seen between the group "with a bachelor's degree or higher" and the other two groups.

Table 11.7 Trust scores of the three groups with different education backgrounds

News type	Credibility evaluation of the three groups with different education backgrounds								Significance (P)
	Undergraduates		Lower than a bachelor's degree		With a bachelor's degree or higher		Overall		
	Sample size	Score	Sample size	Score	Sample size	Score	Sample size	Score	
Sports	214	8.26	188	8.00	223	8.06	625	8.11	0.252
World	218	7.54	180	7.84	235	7.54	633	7.63	0.175
Life	206	7.57	156	7.21	217	7.28	579	7.36	0.132
Legal	207	7.37	185	7.58	217	6.89	609	7.27	0.001
Business	207	7.51	171	7.22	225	6.82	603	7.17	0.000
Social news	209	7.18	179	7.46	222	6.57	610	7.04	0.000
Disaster and accidents	225	7.02	204	7.47	244	6.33	673	6.91	0.000
Domestic politics	217	7.29	172	7.08	234	6.31	623	6.86	0.000
Public safety	220	6.61	201	6.9	242	5.78	663	6.39	0.000
Entertainment	215	5.94	178	6.63	228	6.09	621	6.19	0.011
Unemployment	222	6.38	200	6.57	245	5.48	667	6.10	0.000
Fighting corruption	224	6.08	203	6.45	246	5.33	673	5.92	0.000
Special Campaign	201	5.98	151	6.42	213	5.29	565	5.84	0.000

11.2.3 The Credibility Evaluation Results of Different Media

11.2.3.1 Assessment Methods and Questionnaire Design

We conducted a credibility survey of 36 media and websites that were popular in Beijing, and a 10-point scoring system was used. For media of same type, the "relative credibility" was evaluated. The questions were designed as followings;

> Generally, how much do you trust the following media? Please use 0–10 points to score according to your feelings. A minimum of 0 point indicates complete distrust, a maximum of 10 points indicates complete trust, and 6 points pass. The more trust, the higher the score. Don't rate the one that you are not familiar with.
>
> How much can you trust the following TV stations?
>
> 1. *CCTV* ( ) 2. *Beijing TV* ( )
> 3. *Hunan TV* ( ) 4. *Dragon TV* ( )
> 5. *Phoenix TV* ( )
>
> Relatively speaking, which the above TV stations do you trust the most?
> How much can you trust the following newspapers?

1. *People's Daily*	( )	2. *Guangming Daily*	( )
3. *Beijing Daily*	( )	4. *Beijing Evening News*	( )
5. *Beijing Times*	( )	6. *Beijing Star Daily*	( )
7.*BeijingMorning News*	( )	8. *The Beijing News*	( )
9. *Beijing Youth Daily*	( )	10. *Southern Weekly*	( )
11. *Reference News*	( )	12. *Global Times*	( )
13. *Life Style*	( )	14. *Titan Sports*	( )
15.*China Business Journal*	( )	16. *The Economic Observer*	( )
17. *Economic Daily*	( )	18. *21st Century Business Herald*	( )

Relatively speaking, which one do you trust the most?
How much can you trust the following magazines?

1. *Caijing*	( )	2. *Life Week*	( )
3. *South Reviews*	( )	4. *Outlook weekly*	( )
5. *China Comment*	( )	6. *Oriental Outlook*	( )
7. *China Newsweek*	( )	8. *Phoenix Weekly*	( )

Relatively speaking, which one do you trust the most?
How much can you trust the news posted on the following websites?

1. *Sina.com*	( )	2. *Sohu.com*	( )
3. *People.com*	( )	4. *Xinhua.com*	( )
5. *Qianlong.com*	( )		

Relatively speaking, which website do you trust the most?

11.2.3.2 A Comprehensive Comparison of the Credibility of 36 Media

Since all the media are scored on a 10-point scale, different types of media can be compared.

(1) *Comparison results*

The results are shown in Table 11.8 (the items are arranged in descending order of the score).

(2) *The credibility evaluation results.*

It can be seen for the overall data that the group with higher education level and younger age ranked credibility of *Phoenix TV* and *Phoenix Weekly* at 1st and 2nd, the *Reference News* at 3rd, and *the CCTV* and *People's Daily* at 8th and 11st respectively.

Four professional newspapers and magazines rank the top 10, namely *Global Times, Titan Sports, The Economic Observer* and *Caijing*. The results are similar to people's evaluation of the credibility of international news, sports news, and financial news. In addition, *Southern Weekly* and *China Newsweek*, which are known for their

Table 11.8 36 media's credibility score, ranking, exposure rate and differences

Media	Score and ranking								Significance (P)	Exposure rate (%)
	Undergraduates		Lower than a bachelor's degree		With a bachelor's degree or higher		Overall			
	Score	Ranking	Score	Ranking	Score	Ranking	Score	Ranking		
Phoenix TV	8.33	1	7.66	9	8.22	1	8.11	1	0.000	79.2
Phoenix Weekly	8.26	2	7.43	15	7.95	3	7.94	2	0.005	31.6
Reference News	8.15	4	7.84	6	7.79	4	7.93	3	0.093	68.5
Southern Weekly	8.11	5	7.04	25	7.99	2	7.85	4	0.000	51.4
Global Times	8.11	6	7.96	3	7.48	7	7.83	5	0.007	49.9
Titan Sports	8.22	3	7.63	10	7.52	6	7.81	6	0.003	44.8
China Newsweek	7.75	10	8.03	2	7.47	8	7.75	7	0.120	30.8
CCTV	7.60	16	8.53	1	6.98	14	7.66	8	0.000	98.8
The Economic Observer	8.04	7	7.40	16	7.59	5	7.65	9	0.132	28.9
Caijing	7.88	8	7.53	11	7.38	9	7.54	10	0.325	30.6
People's Daily	7.74	12	7.89	5	6.74	20	7.40	11	0.000	73.4
Life Week	7.58	17	7.49	12	7.21	12	7.39	12	0.416	22.7
China Comment	7.69	14	7.38	17	7.07	13	7.39	13	0.037	44.9
South Reviews	7.81	9	6.43	34	7.36	10	7.35	14	0.000	26.2
Beijing Youth Daily	7.68	15	7.48	14	6.94	16	7.32	15	0.002	56.8
Outlook weekly	7.75	11	6.98	26	6.79	18	7.21	16	0.003	22.3
21st Century Business Herald	7.72	13	6.59	33	7.29	11	7.20	17	0.026	26.3
Beijing TV	7.04	29	7.91	4	6.59	27	7.18	18	0.000	79.8
Beijing Evening News	7.27	24	7.84	7	6.39	30	7.16	19	0.000	54.5
Xinhua.com	7.48	20	7.25	20	6.84	17	7.15	20	0.036	48.0
Beijing Times	7.02	31	7.74	8	6.50	28	7.13	21	0.000	56.9
Sohu.com	7.33	21	7.34	18	6.77	19	7.11	22	0.001	78.8
Beijing Daily	7.51	19	7.48	13	6.13	33	7.02	23	0.000	43.9
The Beijing News	7.19	27	7.19	22	6.73	21	7.01	24	0.021	59.5
Guangming Daily	7.31	22	6.77	29	6.68	22	6.95	25	0.110	40.2
Economic Daily	7.56	18	6.80	28	6.63	25	6.92	26	0.067	26.4
Beijing Morning News	6.96	32	7.33	19	6.42	29	6.92	27	0.001	40.6
Oriental Outlook	7.22	26	6.69	31	6.94	15	6.91	28	0.544	13.6
Sina.com	7.04	30	7.19	21	6.64	24	6.91	29	0.014	82.9
Dragon TV	7.25	25	6.84	27	6.62	26	6.90	30	0.022	57.9
People.com	7.31	23	7.13	23	6.25	31	6.84	31	0.002	41.0
Hunan TV	6.93	33	7.12	24	6.22	32	6.73	32	0.000	70.7
China Business Journal	7.11	28	6.62	32	6.65	23	6.72	33	0.502	22.1
Qianlong.com	6.58	34	6.31	36	6.04	34	6.28	34	0.351	27.2
Life Style	6.53	35	6.31	35	5.85	35	6.18	35	0.080	40.6
Beijing Star Daily	6.28	36	6.72	30	5.32	36	6.08	36	0.000	36.1

in-depth comments, rank the top 10. There is none of the daily newspapers in top 10, and *People's Daily* (score 7.40, ranking 11th) is the most credible newspaper. Although *The Beijing News* is intended to be the mainstream daily newspaper, its exposure rate is 59.5%, with a credibility score of 7.01, ranking 24th.

Since respondents were required not to give score to the unfamiliar media, the exposure rate can be calculated. The results do not show the direct relation between media exposure rate and media's credibility. Four of the top 10 media have the exposure rate of only about 30%. The media with the exposure rate higher than 70%, *Beijing TV* and *Hunan TV*, rank at 18th and 32nd respectively.

(3) *Comparison of the results of groups with different education backgrounds*

There is a significant difference among the results of groups with different education backgrounds. The top 11 is listed in Table 11.8. For the convenience of analysis, they are divided into "political orientation", "professional orientation" and "popular orientation" according to their characteristics. The groups' evaluations are quite different in the three types. It reveals the divergence in the evaluation criteria of media credibility. Therefore, these three categories represent the characteristic of the media itself, as well as different methods of evaluation (see Table 11.9).

Table 11.9 reveals a considerable difference in the evaluation by groups with different education backgrounds, which are not merely the differences in degree, but also in the concept of credibility evaluation. The analysis in Chap. 9 has shown that groups with different education backgrounds have different or even opposing evaluation criteria. The results from the table above support this conclusion.

The media's political importance generally means that the media is run by the higher-level Party or government departments, which can be deemed as their spokesperson. Such media has a high political authority and relies on the influence

Table 11.9 Media with significance in evaluation by groups with different education backgrounds

Item	Media	Score and ranking						Significance (P)	Exposure rate (%)
		Undergraduates		Lower than a bachelor's degree		With a bachelor's degree or higher			
		Score	Ranking	Score	Ranking	Score	Ranking		
Political orientation	*CCTV*	7.60	16	8.53	1	6.98	14	0.000	98.8
	People's Daily	7.74	12	7.89	5	6.74	20	0.000	73.4
	Beijing Daily	7.51	19	7.48	13	6.13	33	0.000	43.9
Professional orientation	*Phoenix TV*	8.33	1	7.66	9	8.22	1	0.000	79.2
	Phoenix Weekly	8.26	2	7.43	15	7.95	3	0.005	31.6
	Southern Weekly	8.11	5	7.04	25	7.99	2	0.000	51.4
	South Reviews	7.81	9	6.43	34	7.36	10	0.000	26.2
Popular orientation	*Beijing Times*	7.02	31	7.74	8	6.50	28	0.000	56.9
	Beijing Evening News	7.27	24	7.84	7	6.39	30	0.000	54.5
	Beijing TV	7.04	29	7.91	4	6.59	27	0.000	79.8
	Hunan TV	6.93	33	7.12	24	6.22	32	0.000	70.7

of its department. Before China's reform and opening up, Chinese people took "political importance" as the main criteria, which can be referred to as "political orientation". The results of this survey demonstrates that many people still hold this view. Among the 36 media listed in this survey, *CCTV* as TV with the highest authority, and *People's Daily* and *Beijing Daily* as the highest-level Party newspapers. They have the most political importance.

The scores of the three media given by the group "lower than a bachelor's degree" are much higher than those of the other two, and the difference from the group "with a bachelor's degree or higher" is particularly obvious. The *CCTV* ranked at 1st by the group "lower than a bachelor's degree" is ranked at 14th by the group "with a bachelor's degree or higher". *People's Daily* is ranked at 5th and 20th respectively. *Beijing Daily* is ranked at 13th and 33rd respectively with a big difference of 20. It can be seen that the evaluation by people "lower than a bachelor's degree" has an obvious "political orientation". It should be pointed out that the political importance is not completely related to the level of the authority. For example, *Guangming Daily* and *People's Daily* are run by the Central Government, but their credibility are quite different.

With the opening up of China's media, a group of media which are targeted at the elites have emerged. They use Western "professionalism" as the media criterion, which can be referred to as "professional oriented" media. The representatives of such media are *Phoenix TV*, *Phoenix Weekly, Southern Weekly* and *South Reviews.*

The audience of such media is generally highly educated, with a wide range of knowledge, and a strong ability to analyse and identify information and independent thinking. Their evaluation criteria are quite different from the traditional "political orientation", so it is referred to as the "professional orientation". It can be seen from the Table 11.9 that the "undergraduates" and the group "with a bachelor's degree or higher" gave a high evaluation of "professional orientation" media, while the evaluation from the group "lower than a bachelor's degree" is less positive. *Southern Weekly* ranked at 2nd by group "with a bachelor's degree or higher" is ranked 25th by the group "lower than a bachelor's degree", which is lower than the *Hunan TV* (24th). *South Reviews* is ranked at 10th and 34th respectively by the two groups; *Phoenix TV* at 1st and 9th respectively, and *Phoenix Weekly at* 3rd and 15th respectively. The evaluation of the group with higher education level has an obvious "professional orientation".

There is another type of media that follows market principles and aims to raise circulation or ratings. It is referred to as a "popular orientation" media, including *Beijing Times, Beijing Evening News, Beijing TV and Hunan TV*. For these media, the evaluation by the group "lower than a bachelor's degree" is significantly higher than that of the other two groups, with a minimum difference of 8 places and a maximum of 25 places. It can be seen that the evaluation of the people "lower than a bachelor's degree" is relatively popular-oriented.

(4) *Comparison of evaluation results and assessment methods*

In this survey the 10-point score system (the "absolute credibility evaluation") and "relative credibility evaluation" are both applied. The relative method requires the

Table 11.10 The credibility evaluation results of 5 TV stations

Media	Relative credibility evaluation		Absolute credibility evaluation			Ranking	
	Sample size	Relative credibility (%)	Sample size	Absolute credibility (%)	Exposure rate (%)	Relative credibility	Absolute credibility
CCTV	274	39.6	684	7.66	98.8	1	2
Phoenix TV	241	34.8	548	8.11	79.2	2	1
Beijing TV	27	3.9	552	7.18	79.8	3	3
Dragon TV	12	1.7	401	6.90	57.9	4	4
Hunan TV	10	1.4	489	6.73	70.7	5	5
Valid sample size	564	81.5					
Missing values	128	18.5					
Total	692	100					

respondents to select a most trusted medium from a certain type of media. The results and ranks from the two methods are listed in Tables 11.10, 11.11, 11.12 and 11.13 (items are arranged in descending order of relative credibility results).

In the result of "relative credibility evaluation", *CCTV* ranks 1st. The previous analysis shows that relative evaluation is influenced by the popularity of the media itself, and it is a combination of multiple variables such as prevalence rate, perception rate and exposure rate. *Phoenix TV* is oversea media, so not everyone has the access to it. The survey shows that *Phoenix TV*'s exposure rate is nearly 20 percentage points lower than that of *CCTV*. If the samples are people who have watched both TVs, there are 446 respondents who meet the standard. The results of "relative credibility evaluation" show that 53.1% of them trust *Phoenix TV* most, which accounts for 34.2% of the total samples. The proportion for *CCTV* is 41.0%, accounting for 26.4%. The rank of *Phoenix TV* has exceeded that of *CCTV*. The difference in the two rankings shows that quite a number of the respondents who chose *CCTV* as the most trusted have only watched *CCTV*, but have not been exposed to *Phoenix TV*.

Results of the "relative credibility" and "absolute credibility" of newspapers, magazines, and websites are shown in Tables 11.11, 11.12 and 11.13. Such differences can be found in them, which indicates the difference in credibility assessment methods can lead to different results. We should be cautious about the assessment methods, especially using reference data from previous researches.

Table 11.11 The credibility evaluation results of 18 newspapers

Media	Relative credibility evaluation		Absolute credibility evaluation			Ranking	
	Sample size	Relative credibility (%)	Sample size	Absolute Credibility (%)	Exposure rate (%)	Relative credibility	Absolute credibility
People's Daily	121	17.5	508	7.40	73.4	1	6
Reference News	108	15.6	474	7.93	68.5	2	1
Southern Weekly	91	13.2	356	7.85	51.4	3	2
Beijing Times	50	7.2	394	7.13	56.9	4	10
Global Times	41	5.9	345	7.83	49.9	5	3
Beijing Youth Daily	33	4.8	393	7.32	56.8	6	7
Beijing Evening News	32	4.6	377	7.16	54.5	7	9
Titan Sports	31	4.5	310	7.81	44.8	8	4
Beijing Star Daily	19	2.7	412	7.01	59.5	9	12
The Economic Observer	16	2.3	200	7.65	28.9	10	5
Beijing Star Daily	12	1.7	250	6.08	36.1	11	18
Beijing Daily	9	1.3	304	7.02	43.9	12	11
Beijing Morning News	8	1.2	281	6.92	40.6	13	15
21st Century Business Herald	6	0.9	182	7.20	26.3	14	8
Guangming Daily	4	0.6	278	6.95	40.2	15	13
Life Style	4	0.6	281	6.18	40.6	16	17
China Business Journal	2	0.3	153	6.72	22.1	17	18

(continued)

Table 11.11 (continued)

Media	Relative credibility evaluation		Absolute credibility evaluation			Ranking	
	Sample size	Relative credibility (%)	Sample size	Absolute Credibility (%)	Exposure rate (%)	Relative credibility	Absolute credibility
Economic Daily	0	0			26.4	18	14
Valid sample size	587	84.8					
Missing values	105	15.2					
Total	692	100					

Table 11.12 The credibility evaluation results of 8 magazines

Media	Relative credibility evaluation		Absolute credibility evaluation			Ranking	
	Sample size	Relative credibility (%)	Sample size	Absolute credibility (%)	Exposure rate (%)	Relative credibility	Absolute credibility
China Comment	83	12.0	311	7.39	44.9	1	5
Phoenix Weekly	80	11.6	219	7.94	31.6	2	1
China Newsweek	64	9.2	213	7.75	30.8	3	2
Caijing	58	8.4	212	7.54	30.6	4	3
South Reviews	45	6.5	181	7.35	26.2	5	6
Life Week	28	4.0	157	7.39	22.7	6	4
Oriental Outlook	15	2.2	94	6.91	13.6	7	8
Outlook weekly	13	1.9	154	7.21	22.3	8	7
Valid sample size	386	55.8					
Missing values	306	44.2					
Total	692	100					

Table 11.13 The credibility evaluation results of 5 websites

Media	Relative credibility evaluation		Absolute credibility evaluation			Ranking	
	Sample size	Relative credibility (%)	Sample size	Absolute credibility (%)	Exposure rate (%)	Relative credibility	Absolute credibility
Sina.com	184	26.6	574	6.91	82.9	1	3
Sohu.com	131	18.9	545	7.11	78.8	2	2
Xinhua.com	113	16.3	332	7.15	48.0	3	1
People.com	59	8.5	284	6.84	41.0	4	4
Qianlong.com	9	1.3	188	6.28	27.2	5	5
Valid sample size	496	71.7					
Missing values	196	28.3					
Total	692	100					

11.3 Summary

This chapter uses different methods to evaluate the credibility of different media. Due to limitations of the sample size, the focus of this chapter is to compare the credibility assessment methods and credibility evaluation of different groups of people. Differences in results of the credibility evaluation of various media reveal the complexity of research methods, and at the same time reflect some potential problems in China's media credibility evaluation. These issues are worth further discussion in future research.

Chapter 12
Conclusion

The empirical research on media credibility is on the basis of quantitative evaluation, and the key to evaluation is to find suitable assessment methods. This research is to explore a proper scale for evaluating credibility of the China's mass media. Although the scale we finally adopt contains only a few simple indexes, the development of a scale to assess the abstract concept of credibility requires not only proper indexes, but also covers the definition of credibility and other issues related to cross-cultural transformation. They are basic elements of credibility evaluation. After solving these problems, the credibility assessment scale can be achieved.

Media credibility research originated in the United States where rich research results including media credibility assessment scale are achieved. However, the research on media credibility in China has just begun. Under this context there are two research methods for evaluating China's media credibility. One is to adopt the American assessment scale to evaluate China's media, and the other starts with reviewing the concept of credibility, following the general steps to reconstruct indexes of the evaluation scale. With an in-depth review of credibility study in the United States, it is found that there are cross-cultural differences in the media credibility evaluation criteria. However, the concept and scale of credibility are not yet perfect, so we do not "pick up ready-made fruits". Instead, we choose to construct the credibility assessment scale. This process is not to deny previous research results but to review and absorb them.

The development of the credibility assessment scale is like a journey of discovery of media credibility, where many valuable research topics are found. This research journey is just beginning.

Firstly, this research reviews the former credibility research and their social contexts, by which three objectives are achieved: (1) to provide valuable information for the definition of media credibility and the development of assessment scales; (2) to examine deficiencies of the current research; (3) to demonstrate that the study of credibility is closely related to specific social and cultural backgrounds, so the applicability of results to cross-cultural research needs to be addressed with deliberation.

G. Yu, *Research on the Communication Effects and Mass Media Credibility in China*,
https://doi.org/10.1007/978-981-19-6242-4_12

Secondly, this study introduces the main propositions, research designs and research methods. The three propositions are analysed in depth: (1) discussion on the concept of credibility; (2) the criteria and characteristics of China's media credibility evaluation; (3) the development of credibility assessment scale.

Thirdly, based on the literature review, related concepts and theories are explored. This research analyses the importance and characteristics of media credibility evaluation criteria with the support of data. It probes into the construction of the scale and develops the preliminary indexes for the assessment scale.

Fourthly, on the basis of the previous research, through statistical techniques, such as the exploratory and confirmatory factor analysis, reliability and validity test and others, indexes of the scale are selected and analysed, and it makes an attempt to construct a media credibility assessment scale.

References

Abel, J. D., & Wirth, M. O. (1977). Newspaper vs. TV credibility for local news. *Journalism Quarterly, 54*(2), 371–375.
Ai, F. (1999). *Public opinion supervision and news planning* (pp. 5–7). Sichuan People's Publishing House.
Almond, G. A., & Powell, G. B., Jr. (1987). *Comparative politics: System, process, and policy* (P. Can et al., Trans., p. 168). Shanghai Translation Publishing House.
Becker, L., Whitney, C., & Collins, E. (1980). Public understanding of how the news media operate. *Journalism Quarterly, 57*(4), 571–578.
Berlo, D. K., Lemert, J. B., & Mertz, R. J. (1969). Dimensions for evaluating the acceptability of message sources. *Public Opinion Quarterly, 33*(4), 563–576.
Bu, W. (2002). *The impact of mass media on children.* Xinhua Publishing House.
Butler, J. (1991). Toward understanding and measuring conditions of trust: Evolution of a conditions of trust inventory. *Journal of Management, 17*(3), 643–663.
Cao, L. (2004). Analysis of the pursuit of Southern Weekly. *Journal of Yili Normal University, 1.*
Cao, R. (2003). *Out of the prisoner's dilemma: An analysis of social capital and institution.* Shanghai SDX Joint Publiy Shing Company.
Carruthers, S. L. (1995). *The media at war* (Z. Zhang et al., Trans.). Xinhua Publishing House.
Carter, R. F., & Greenberg, B. S. (1965). Newspapers or television: Which do you believe? *Journalism Quarterly, 42*(1), 29–34.
Chen, G., & Chun, T. (2004). *Survey of Chinese farmers.* People's Literature Publishing House.
Chen, H. (1999a). The evolution of the Chinese media system in the 1990s. *Twenty-First Century, 53.*
Chen, L. (1999b). *Public opinion—Study on opinion orientation.* China Radio Film & TV Press.
Chen, S. (1983). *Mass communication and social change* (pp. 56–58). Taiwan Sanmin Bookstore.
Chen, X. (2003). *China internet media forum: Credibility—Lifeline of Internet development.* Retrieved May 10, 2020, from http://www.cctv.com/tvguide/tvcomment/wtjj/xzlz/7517.shtml
Chen, X. (2004). The constituent elements of media credibility. *Press Outpost, 5.*
Cui, H. (2004, September 23). Government agencies in Beijing will regularly publicize major administrative policies. *Beijing Morning Post.* Retrieved May 15, 2020, from http://news.sina.com.cn
Dasgupta, P. (2000). Trust as a commodity. In D. Gambetta (Ed.), *Trust: Making and breaking cooperative relations* (pp. 49–72). University of Oxford.
Davis, S., & Lawrence, P. R. (1977). *Matrix.* Addison-Wesley.
Defleur, M. L., & Ball-Rokeach, S. (1990). *Theories of mass communication* (L. Du, Trans.). Xinhua Publishing House.
Ding, B., & Nan, H. (2004). Authority, profundity, and practicableness—On the core value of the economic report of the Party newspaper. *News Practice, 7.*

G. Yu, *Research on the Communication Effects and Mass Media Credibility in China*,
https://doi.org/10.1007/978-981-19-6242-4

Ding, S. (2003). Credibility is a valuable asset of the media. *Lianhe Zaobao*, 2003-03-15.

Donn, J. (2003). Trust and political agency. In Y. Zheng (Ed.), *Trust: Making and breaking cooperative relations* (p. 89). China City Press.

Du, P. (2004). The media cannot make waves in the world any longer. *Lianhe Zaobao*. Retrieved May 22, 2020, from http://www.zaobao.com/yl/tx501_221004.html

Dube, R. (1998). *Focus on attention: A behavioral perspective on media credibility*. Doctor dissertation, Washington University.

Edelstein, A. S., Ito, Y., & Kepplinger, H. M. (1989). *Communication & culture: A comparative approach*. Longman.

Fang, L. (1996). *Principle of communication*. Taiwan Sanmin Bookstore.

Feng, Y., & Yu, Q. (2004). Analysis of the crisis and construction of the creditability of market newspapers. *Hubei Social Science, 8*.

Fogg, B. J., et al. (2002). *Stanford-Makovsky web credibility study 2002: Investigating what makes web sites credible today*. A Research Report by the Stanford Persuasive Technology Lab & Makovsky & Company. Stanford University. www.webcredibility.org

Fukuyama, F. (2001). *Trust: The social virtues and the creation of prosperity* (Z. Peng, Trans.). Hainan Publishing House.

Gabarro, J. (1987). *The dynamics of taking charge*. Harvard Business School Press.

Gambetta, D. (2003). Can we trust trust? In Y. Zheng (Ed.), *Trust: Making and breaking cooperative relations* (Y. Yang & S. Peng, Trans., pp. 270–271). China City Press.

Ganahl, R. (1994). *Newspaper readership and credibility: An application of media uses and gratification theory*. Unpublished doctoral dissertation, University of Missouri, Columbia.

Gaziano, C., & McGrath, K. (1986). Measuring the concept of credibility. *Journalism Quarterly, 63*(3), 451–462.

Gellner, E. (2003). Trust, cohesion and the social order. In Y. Zheng (Ed.), *Trust: Making and breaking cooperative relations* (pp. 177–178). China City Press.

Giddens, A. (1991). *Modernity and self-identity: Self and society in the late modern age* (p. 4). Stanford University Press.

Gong, W. (1994). The origin and development of Persuasion—From Aristotle's ethos, pathos and logos. *Journal of Beijing University* (Social Science Edition), *3*.

Good, D. (2003). Individual, interpersonal relationship, and trust. In Y. Zheng (Ed.), *Trust: Making and breaking cooperative relations*. China City Press.

Gunther, A. C. (1992). Biased press or biased public? Attitudes toward media coverage of social groups. *Public Opinion Quarterly, 56*, 147–167.

Guo, Q. (1999a). *An introduction to communication*. China Renmin University Press.

Guo, S. (2003, August 5). Internet credibility ranked second. *Beijing Morning Post*. Retrieved May 15, 2020, from http://news.sina.com.cn

Guo, Z. (1999b). *Social statistical analysis methods—Application of SPSS software* (p. 351). China Renmin University Press.

Hart, K. (2003). Kinship, contract, and trust: The economic organization of migrants in African city slum. In Y. Zheng (Ed.), *Trust: Making and breaking cooperative relations* (p. 232). China City Press.

He, G. (2004). Analysis of the mass media credibility formation process. *Press Circles, 2*.

He, S., & Wang, S. (2002). A system analysis of the correlation between the local government's trustworthiness and the government's operating costs. *Journal of the National Academy of Governance* (Special Issue).

Hovland, C. L., Irving, L. J., & Harold, H. K. (1953). *Communication and persuasion*. Yale University Press.

Huang; X. (1999). Credibility and the authority of the media. *TV Research, 11*.

Izard, R. (1985). Public confidence in the news media. *Journalism Quarterly, 62*, 247–255.

Jacobson, H. K. (1969). Mass media believability. A study of receiver judgments. *Journalism Quarterly, 46*(1), 20–28.

Johnson, T. J., & Kaye, B. K. (1998). Cruising is believing? Comparing internet and traditional sources on media credibility measures. *Journalism Quarterly, 75*(2), 523–540.

Ke, H., Zhu, J., & Sun, J. (2003). *Communication statistics* (p. 376). Beijing Broadcast College Press.

Kiousis, S. (2001). Publictrust or mistrust? Perceptions of media credibility in the information age. *Mass Communication & Society, 4*(4), 381–403.

Koehn, D. (2003). *Trust and business: Barriers and bridges* (X. Lu, Trans.). Shanghai Academy of Social Sciences Press.

Lee, R. S. H. (1978). Credibility of newspaper and TV news. *Journalism Quarterly, 55*(2), 282–287.

Legal Daily. (2003, May 28). *Reflection on SARS: Information is open and why rumors are still believed*. Retrieved May 15, 2020, from http://news.sina.com.cn

Lewicki, R. J., & Barbara, B. B. (2003). Developing and maintaining trust in work relationships. In Kramer & Tyler (Eds.), *Trust in organizations: Frontiers of theory and research*. China City Press.

Li, H., & Yang, X. (2000). *Social capital and social development*. Social Science Literature Press.

Li, J. (2003). Extraordinary events and the media credibility. *Press Outpost, 4*.

Li, J. (2004). Problematic reports demonstrating the media credibility. *Media Observer, 8*.

Li, J. (1996). *Theory of mass communication* (p. 18). Taiwan Sanmin Bookstore.

Li, L. (1997). *An introduction to western journalism* (p. 85). Fudan University Press.

Li, M. (1994). *A new discussion on mass communication*. Taiwan Sanmin Bookstore.

Li, X., & Qin, X. (2001). Who is setting China's agenda today?—E-forum reconstruction of party newspaper in major news events. *Journal of Journalism and Communication Studies, 3*.

Li, Y. (2005). Truth is in the hands of reporters? Internal evaluation mechanism and distortion of news in market-based media. *Youth Journalist, 1*.

Li, Z. (2003). On the mass media credibility. *Journal of Xi'an University of Architecture and Technology* (Social Science Edition), *1*.

Lin, A., & Wang, X. (2003). Facing the contemporary audience: Media literacy education. *Science News, 24*, 19–22.

Lin, H. (2004a). *Why the power of CCTV public opinion supervision is great?* Retrieved May 15, 2020, from http://www.hi.chinanews.com/hnnew/2004-05-29/5507.html

Lin, Y. (2004b). Media use and social construction: A preliminary study of the relations between Macau's media and social change. In *The thesis collection of the 2004 National Ph.D. Discussion Forum-Social Science Disciplines at Renmin University of China.*

Lin, Z. (2004c). *New approach to communication psychology* (p. 146). Peking University Press.

Littlejohn, W. S. (1999). *Theories of human communication* (D. Chen & X. Ye, Trans.). China Social Sciences Press.

Liu, J. (1997). *Psychology of mass communication* (p. 137). Beijing Broadcast College Press.

Liu, J. (1998). *The course of nature and the will of the people—The social opinion of contemporary China*. China Today Publishing House.

Liu, X. (2003). *China internet media forum: Credibility—Lifeline of internet development*. Retrieved May 10, 2020, from http://www.cctv.com/tvguide/tvcomment/wtjj/xzlz/7517.shtml

Lo, V. (1993). *Journalism theory and evidence* (pp. 12–16). Liming Cultural Corporation.

Lo, V., & Chen, S. (1993). Research on the credibility of the news media. *Special Research Report of the Science Council*.

Lo, V., Chen, T., & Pan, Z. (2001). Ethical attitudes and perceived practice: A comparative study of journalists in China's mainland, Hong Kong, and Taiwan. *Journalism Studies, 68*.

Lo, V., Lin. W., Niu, L., & Tsai, C. (2003). The impact of media reliance and media use on election news credibility of: A comparison of five media. *Journalism Research, 74*.

Luo, J. (2003, January 18). *A speech at the academic seminar of "Media Change and Aesthetic Cultural Innovation"*. Beijing Normal University.

Lu, Y., & Wang, Y. (2000). *Mass culture and media*. Shanghai SDX Joint Publishing Company.

Lu, Y., & Pan, Z. (2002). The imagination of fame. *Journalism Research, 4*.

Lu, Y., & Pan, Z. (2004). Imaging professional fame: Constructing journalistic professionalism in China's social transformation. *Mass Communication Research, 4*, 17–59.

Ma, Q., & Du, Z. (2001). Using credit to drive the market economy. *Hubei Social Sciences, 3.*

Markham, D. (1968). The dimensions of source credibility of television newscasters. *Journal of Communication, 18*, 57–64.

McCroskey, J. C., & Jenson, T. A. (1975). Image of mass media news sources. *Journal of Broadcasting, 19*, 169–180.

McGregor, D. (1967). *The professional manager*. McGraw-Hill.

McQuail, D. (1994). *Mass communication theory: An introduction.* Sage.

McQuail, D., & Windahl, S. (1997). *Communication models for the study of mass communications* (J. Zhu & W. Wu, Trans., pp. 47–50). Shanghai Translation Publishing House.

Mencher, M. (2003). *News reporting and writing* (J. Zhan, Trans., p. 70). Huaxia Publishing House.

Meyer, P. (1988). Defining and measuring credibility of newspapers: Developing an index. *Journalism Quarterly, 65*(3), 567–574.

Mishra, A. (2003). Organizational responses to crisis—The centrality of trust. In R. M. Kramer & T. R. Taylor (Eds.), *Trust in organizations.* China City Press.

Mrogers, E., & Wdearing, J. (2000). Agenda-setting research: Where has it been, where is it going? In *Mass communication: The paradigm of impact studies* (S. Guan et al., Trans., p. 83). China Social Sciences Press.

Mulder, R. A. (1981). A log-linear analysis of media credibility. *Journalism Quarterly, 58*(4), 635–638.

Newhagen, J., & Nass, C. (1989). Differential criteria for evaluating credibility of newspapers and TV News. *Journalism Quarterly, 66*(2), 277–284.

North, D. (1994). *Structure and change in economic history* (Y. Chen & H. Luo, Trans.). Shanghai SDX Joint Publishing Company.

Orum, A. M. (1989). *Introduction to political sociology: A social analysis of political entities* (Y. Dong & Y. Li, Trans.). Zhejiang People's Publishing House.

Ouchi, W. G. (1981). *Theory Z: How American business can meet the Japanese challenge.* Addison-Wesley.

Pagden, A. (2000). The destruction of trust and its economic consequences in the case of eighteenth-century Naples. In D. Gambetta (Ed.), *Trust: Making and breaking cooperative relations* (electronic ed., pp. 127–141) University of Oxford.

Pan, J., Wang, S., & Xie, Y. (1990). Research on mass communication and issue setting in the Taiwan region. *Special Research Report of the Science Council.*

Pang, J. (2004, April 3). Chengdu will start freely publicizing government regulations and people's livelihood information on May 1. *Chengdu Economic Daily.* Retrieved May 15, 2020, http://news.sina.com.cn

Peng, S. (1999). Trust-building mechanism: Relationship and legal means. *Sociological Studies, 2.*

Peng, Y. (1991). The evaluation of the media's credibility during the election by the media directors. *Journalism Research, 44.*

Putnam, H. (1981). *Reason, truth and history* (p. 128). Cambridge University Press.

Richardson, B., Detweiler, J. S., & Bush, M. B. (1988). *Linkages between journalists' community associations, attitudes and expression of viewpoints on selected issues.* Paper presented to Association for Education in Journalism and Communication, Portland, Ore.

Rimmer, T., & Weaver, D. (1987). Different questions, different answers? Media use and Media credibility. *Journalism Quarterly, 64*(1), 28–36.

Schramm, W. (1990). *Mass media and national development: The role of information in the developing countries* (Y. Jin et al., Trans., p. 150). Huaxia Publishing House.

Schweiger, W. (2000). Media credibility—Experience or image? A survey on the credibility of the World Wide Web in Germany in comparison to other media. *European Journal of Communication, 15*(1), 37–59.

Shang, D., & Ke, H. (2002). Impact of various Chinese audience types on empirical research in the social transition. *Modern Communication, 4.*

Shaw, E. F. (1973). Media credibility: Taking a measure of a measure. *Journalism Quarterly, 50*(2), 306–311.

She, W. (2002). Creditability: An important weight for media competition. *News Front, 5*, 32–33.

Singletary, M. W. (1976). Components of credibility of a favorable news source. *Journalism Quarterly, 53*(2), 316–319.

Slater, M., & Rouner, D. (1996). How massager evaluation and source attributes may influence credibility assessment and change. *Journalism & Mass Communication Quarterly, 73*(4), 974–991.

Song, X. (2003). Analysis of honesty and credit. *People's Daily*, May 20.

Stamm, K., & Dube, R. (1994). The relationship of altitudinal components to trust in media. *Communication Research, 21*(1), 105–123.

Sun, L. (2002). Rebuilding social credibility. *The Economic Observer*, April 22.

Sun, Y., & Bu, W. (1998). *Developing healthy personality for only children.* Tianjin Education Press.

Tian, W. (2001). Culture and right—Understanding the cultural orientation in the study of contemporary western political communication theory. *The Social Science Front, 2.*

Time Mirror. (1986). *The People & Press: A time mirror investigation of public attitudes toward the news media.* Conducted by Gallup in collaboration with Michael J. Robertson. Times Mirror Company.

Wang, G. (2004). Basic approaches for media to serve the public. *Journalism Lover, 7.*

Wang, S., & Liu, X. (2003). The base of trust: A rational explanation. In Y. Zheng (Ed.), *Trust in Chinese society* (p. 226).

Wang, X., & Qin, H. (1995). Why there is a crist of trust in TV sales direct. *Market Analysis, 5.*

Wang, X., & Zhao, H. (2004). Investigation on the current status of China's media credibility. *Press Outpost, 4.*

Wang, X., Mo, W., & Tang, Y. (1999). *Media performance: Discussion and assessment of news credibility.* Paper presented to the 1999 Annual Conference of the Chinese Communication Society.

Wanta, W., & Hu, Y. (1994). The effects of credibility, reliance, and exposure on media agenda-setting: A pat analysis model. *Journalism Quarterly, 71*(1), 90–98.

Waters, M. (2000). *Modern sociological theory* (S. Yang, Trans.). Huaxia Publishing House.

Wei, X., & Bo, Y. (2003). *Crisis of credibility: Mapping a serious social problem in China* (p. 225). China Social Sciences Press.

West, M. D. (1994). Validating a scale for the measurement of credibility: A covariance structure modeling approach. *Journalism Quarterly, 71*(1), 159–168.

Westley, B. H., & Severin, W. J. (1964). Some correlates of media credibility. *Journalism Quarterly, 41*(Summer), 325–335.

Wu, J. (2005). *No credit makes the endless harm—Wu Jinglian on credit construction.* Retrieved May 10, 2020, from http://www.people.com.cn

Ye, H. (2000). *A preliminary study on the credibility of network media and its factors: The case of internet users in the Taiwan region.* Master degree thesis, Chung Cheng University.

Yu, G. (2003). *Media influence.* Nanfang Daily Publishing House.

Yu, G., et al. (2003a). The choice of communication channels in the face of important events. *The Journalist*, 6.

Yu, G., et al. (2003b). The communication effect of information transparency processing—Public opinion survey and analysis of the SARS incident. *The Journalist, 7.*

Yu, H. (2004). Viewing credit for the perspective of sociology. *Wen Wei Bao*, 2004-03-03.

Zang, G. (1999). *News media and sources—A discussion on media frame and truth construction.* Taiwan Sanmin Bookstore.

Zhang, G. (2001a). *News media and society* (p. 59). Shanghai People's Publishing House.

Zhang, H. (2001b). Discussion on the relationship between agenda setting theory and frame theory of mass communication. *Journal of Southwest Minzu University* (Social Science Edition), *10.*

Zhang, H. (2004). News security: Thinking based on reality. *Journal of Western Radio and Television, 3.*

Zhang, J. (1995). On election news reporting from the perspective of public domain theory and diversified reporting views. *Journalism Research, 55.*

Zhang, J., & Qiao, S. (2004). The comprehension and practitioner of professionalism in journalism. *China Economic Times*, 2004-10-27.

Zhang, K. (2003a). *The political socialization function of the mass media* (p. 140). Wuhan University Press.

Zhang, W. (2001c). Economists review of law, culture and history. *Global Management Review, 3*, 4–6.

Zhang, W. (2003b). *Information, trust and law* (pp. 35–38). SDX Joint Publishing Company.

Zhang, Z. (2004). Interpretation of the news media credibility. *News Collection, 2.*

Zheng, B. (2003a). *China internet media forum: Credibility—Lifeline of internet development.* Retrieved May 10, 2020, from http://www.cctv.com/tvguide/tvcomment/wtjj/xzlz/7517.shtml

Zheng, B., & Tang, Y. (2004). On the credibility of news media. *Journalism Lover, 3.*

Zheng, Y. (2001). *Trust theory.* China Radio Film & TV Press.

Zheng, Y. (2003b). *Trust: Making and breaking cooperative relations* (p. 365). China City Press.

Zhong, W. (1992). *From media reality to subjective reality.* Cheng Chung Bookstore.

Zhou, S. (2004a). *Concept, psychology, social function and research of media credibility.* One of the achievements of the "Chinese Mass Media Credibility and Communication Effect" project of the University of Alabama, School of Communication and Information Science.

Zhou, X. (1997). *Modern social psychology* (pp. 170–176). Shanghai People's Publishing House.

Zhu, H. (2005). A king of grain: Wandering on the farmland. *Southern Weekly*, 2005-03-10.

Zhu, Q., & Zhu, Y. (2004). On CBS *60 Minutes. Xinhuanet*, 2004-06-14.